GREEN POWER

THE ENVIRONMENT MOVEMENT IN AUSTRALIA

Timothy Doyle

UNSW PRESS

A UNSW Press book

Published by
University of New South Wales Press Ltd
University of New South Wales
UNSW Sydney NSW 2052 Australia
www.unswpress.com.au

© Timothy Doyle
First published 2000
Reprinted with amendments 2001

National Library of Australia
Cataloguing-in-Publication entry:

   Doyle, T.J. (Timothy J.), 1960– .
   Green power: the environment movement in Australia.

   Bibliography.
   Includes index.
   ISBN 0 86840 714 3.

   1.Green movement — Australia. 2. Conservationists —
   Australia — Political activity. 3. Conservation of natural
   resources — Australia. I. Title.

333.720994

Printer BPA, Melbourne

CONTENTS

ACKNOWLEDGEMENTS

Sister Borgias, Bob Boyd, Seamus Doyle, Graeme Duncan, Peter Hay, Austin Ivy, Brian Lewis, Brian Martin, Doug McEachern, Trevor Steele, Ken Walker, Maureen Warburton, Len Webb, Jim Winter and Arnold Zable all deserve mention as my life's teachers.

Thanks are also due to Brian Cook at Clark University in Massachusetts for providing me with the time and office space to write the first few chapters of this book. I must also recognise the University of Adelaide for supplying me with a Small University Research Grant to allow me to pursue the initial literature searches. Thanks to Adam Simpson for his efforts as research assistant in these early stages and to my youngest brother Malachi and Sue Hurst for their editing work.

I am indebted to the staff and students of the Mawson Graduate Centre for Environmental Studies (now the Department of Geographical and Environmental Studies) at the University of Adelaide for their unstinted support, most particularly departmental head Nick Harvey.

I am extremely grateful for the assistance of my publisher, John Elliot, of UNSW Press. His instructive insights given at a crucial stage in the manuscript's development were essential. Many thanks also to my editor, Richard McGregor, for his gentle chiding and thoroughness in the book's final preparation. I am also grateful to Carol Bacchi and Greg McCarthy from the Politics Department at the University of Adelaide for their useful information and encouragement in the final stages.

I must also express my gratitude to those incredible Australians who make up our environment movement: past, present and future. I make special mention of the Conservation Council of South Australia, particularly Michelle Grady, my colleague and executive officer.

Thanks again to both John Elliot and Ken Walker but for different reasons this time: for allowing me to reuse my 'wise use' theme and to further develop some supporting case-study material in chapter 11. I first published part of this material in *Australian Environmental Policy 2* (Walker and Crowley, UNSW Press, 1999).

Similar thanks are accorded to Elizabeth Gibson at Macmillan and to my former co-author Aynsley Kellow for allowing me to adapt a small section from an earlier book of mine, *Environmental Politics and Policy Making in Australia* (Macmillan, 1995), for use in the first part of chapter 2. Most importantly, thanks, as always, to my partner in life, Fiona.

I wish to dedicate this book to my eldest daughter, Georgie. Her courage, her creativity, her compassion for others, her wide-eyed interest in life, and her love and gentleness are inspiration enough for any father to write what is, ultimately, a book of hope.

LIST OF ABBREVIATIONS

ACF	Australian Conservation Foundation
CAFNEC	Cairns and Far North Environment Centre
CCSA	Conservation Council of South Australia
CFND	Campaign for Nuclear Disarmament
EAAAC?!	EcoAnarchoAbsurdistAdelaideCell?!
EF!	Earth First!
EPBC	Environmental Protection and Biodiversity Conservation
ESD	ecologically sustainable development
FIDO	Fraser Island Defenders Organisation
FoE	Friends of the Earth
HSI	International Humane Society
JAG	Jabiluka Action Group
LPAC	The Lake Pedder Action Committee
MDA	militant direct action
NAG	Nomadic Action Group
NDP	Nuclear Disarmament Party
NECF	National Environmental Consultative Forum
NGO	non-governmental organisations
NHT	National Heritage Trust
NVA	non-violent action
QCC	Queensland Conservation Council
RCSQ	Rainforest Conservation Society of Queensland
TCT	Tasmanian Conservation Trust
TRCC	Townsville Rainforest Conservation Council
TWS	The Wilderness Society
UTG	United Tasmania Group
WPSQ	Wildlife Preservation Society of Queensland
WWF	World Wide Fund for Nature

PREFACE

In October 1998 I was fiercely writing this book between the numerous other commitments that make some of our lives a battleground. I was trying to relax in front of the television when *Foreign Correspondent* came on. The image of Australian Broadcasting Corporation (ABC) journalist Evan Williams appeared on the box, all the way from the island of Mindanao in the southern Philippines. He was reporting on both the current and potential impacts of Western Mining Corporation's (WMC) operations at Tampakan, reportedly the second biggest copper deposit in Asia. WMC, a large Australian transnational mining company (predominantly uranium and copper), is expanding its enterprises into a number of Third World economies. Its major mine at Roxby Downs in South Australia is fast becoming one of the largest uranium producers in the world.

Williams told a story under the subtitle 'Another Bougainville in the making'. He argued that WMC, propped up by the Philippines government and military, was challenging such key issues as traditional land ownership of the B'laan people, one of the indigenous peoples of the region. Two things initially came out of the story for me: first, there was something about the beauty of the faces of the B'laan people who were interviewed, and their plight, which jumped out of the screen and physically shook me out of my feeble attempt at relaxation; and second, I understood, for the first time, how similar WMC's operational tactics were in the Philippines when compared with what I had witnessed of their operations in central Australia. In fact, the similarities, I thought, made up a pattern. A seed planted itself in me at a most inconvenient time! I realised that I had insights and information that the Filipino people could use in their battle against the excesses of this company and, indeed, I was sure to learn something of strategic worth for my own organisation, the Conservation Council of South Australia (CCSA), and the broader environment movement, which dealt daily with the antisocial/environmental consequences of WMC's rapid and largely self-monitored expansion of its Australian mining practices.

Within 24 hours I began the task of tracking, mapping and creating networks that would lead me to the coordinators of the campaign against WMC's mine in Mindanao, while respecting cultural protocols. Within three weeks I was on a plane to Manila, and then on to Mindanao, the 'garden' and 'rice bowl' of the Philippines. By coincidence, during this period of planning I was asked, as President of CCSA, to join an International Fact Finding Mission (IFFM) into WMC's operations, led by Pastor Avel Sichon of the Uniting Church of the Philippines. The IFFM included members of church, community and environmental non-governmental organisations from the Philippines and other nations, including one other representative from Australia, Sarah Wright of the Minerals Policy Institute.

At Manila airport, customs officials quizzed us about our purpose in Mindanao, which was regarded as 'unsafe' for tourists. Local newspapers were full of stories of anarchy and a society run by bandits in this southernmost island of the archipelago. Other official reports focused on the formation of large armies in the mountains, such as the Moro Islamic Liberation Front (MILF), the New People's Army, the National Defence Front, the BHP (the soldiers of the people), indigenous tribal warriors such as the Bagani (the warriors of B'laan tribes), and convergences of other militant people's organisations such as the Alliance for Genuine Development.

From General Santos City, on the southern edge of Mindanao, we began to travel by road into the interior of the large and mountainous island. At various stages we conducted extensive interviews with different individuals and groups, including church leaders (both Christian and Muslim), municipal mayors, Filipino 'settler' farming groups, environmental and human rights organisations, teachers in WMC-funded schools, WMC-funded 'community organisers', and WMC employees at the Tampakan base camp.

One attempt to seek this community input will live with me always. The IFFM was camping in the parish compound of a small town at the foot of the 'Mossy Mountain' in Columbo in the province of Sultan Kumarat. Early in the morning we set off on foot to speak to a community of B'laan people living at approximately 1000 metres above sea level. This particular community had refused to sign Western Mining Corporation's Memorandum of Agreement (MoA), which basically entailed handing over all legal and commercial rights of their ancestral domain to the company. When we finally arrived in this tiny village set on the edge of the forest, at the farthest point of assault by global markets, it appeared no-one was there. Pastor Avel, the B'laan priest, told us to sit down, while he moved quietly into the village to investigate and, as we found out later, to negotiate. After a long wait, a cry or whistle rent the air, with warriors and other local people appearing from the forest into the clearing, flooding the village, welcoming us with food and water.

Soon we were sitting in the central hut talking to the elders and warriors of the community. We learned that the people had hidden in the forest when they had seen the colour of my and my countrywoman's skin. Also, with traditional communication, they had heard that Australians were among the party, and Australians could not be trusted as they were the oppressor. I explained that not all Australians supported WMC and other transnational companies mining the Philippines, and how many Australians, including its indigenous peoples and members of the predominantly white environment movement, fought a similar campaign against the company in Australia. I explained that I was here to share these experiences, and to pledge my own organisation's solidarity with their struggle. I told them that they were not alone.

After this initial period, through the translations of our indigenous priest, the IFFM learned, first hand, of the tribe's dealings with WMC. We learned of the specifics of the company's divide-and-rule tactic, the tensions it had created in the community, and the deaths and injuries that had ensued. We learned of the story of one particular B'laan chief, Afnelu-Timon, of Buluf Salo, Kiblanan. The 75th Infantry Regiment had recently burnt all the houses in his village, and killed his people's carabou. The chief believed that this was an act of retribution in response to his tribal group's unwillingness to sign agreements with WMC. This story reinforced Evan Williams' contention that WMC had condoned the use of the military to resolve land conflicts delaying its mining operations.

We discussed strategy. The B'laan explained that they had nowhere to go, that over recent generations, due to different pressures of invasion and resettlement, they had been driven from the lowlands into the highlands. First came the Spaniards, then came the Americans, and now the Australians (or more accurately transnational capital). If they lost this place, then they were dead. They argued against dealing further with the company, as any dealings had always led to trouble and grief. The Bagani warriors sitting quietly in the room (who are arrested and usually shot by the military on sight) explained that they had performed a secret pact known as 'Dyandi'. This ritual relates to the defence of the land 'to the last drop of blood': there was nothing postmaterialist or postmodern about this conflict.[1] It was not about such 'Northern' concepts as biodiversity or greenhouse gas emissions: it was about immediate life or death.

The rain fell in the afternoon, and we were forced to say our good-byes to this particular community in a hurry. The clearance of the forest on the 'Mossy Mountain' by a Japanese logging company had been so intense and so rapid that any significant downpour became a flash flood, and rivers we had crossed at ankle depth on the way up would become walls of water on the way down within hours. As we left, one old B'laan man said good-bye to me through a translator and asked me, in song, a parting question: 'Why has God made you so powerful and the B'laan so powerless?'

That day we finished by literally sliding down the mountain until we came to another small B'laan village near the bottom of the mountain. We were immediately welcomed, and sat and shared food. Although we were welcomed, we had arrived at an inopportune time. Another child had died the night before of fever, and her mother and father were just setting out for a burial in the mountains. We learned of over 20 infant deaths from this one village in the last six months. One of the members of the IFFM was a Filipino specialist in child health and welfare. I asked her whether these deaths were related to dengue fever. She explained that no-one knew, but the Filipinos referred to it as 'the disease of poverty'. She explained that access to something like aspirin would probably have saved the child, but the cause of sickness could not be treated so easily.

For anyone from a predominantly 'Northern' country such as Australia, to truly enter societies of the Third World outside major cities, tourist locations and organised tours is always a culture shock, no matter how experienced, well read and well travelled. The comparative level of poverty just hits you every time. Of course instances of Third World poverty also occur in Australia, in numerous Aboriginal communities and in places where the forgotten though growing underclasses gather and coalesce.

The findings of the IFFM attracted tremendous media coverage in the Philippines. The IFFM told the story that WMC was like McDonald's. As a transnational, it operated to a recipe. And the recipe included two crucial ingredients: first, to deliberately confuse issues of ancestral domain, which led to the disempowerment and dispossession of indigenous peoples; and, second, actively to assimilate and incorporate all opposition into its own management processes. Let me address these two points briefly.

The IFFM discovered, for example, that WMC had employed an anthropologist, Stephen Davis, to provide the initial analysis of ancestral domain issues. WMC had implemented a 'divide and rule' strategy at every level of local society, from the tribal to the municipal, to obtain key signatures on memorandums of agreement, which ultimately allows them total access to ancestral domains. They did this through a series of payments to individuals who agreed with their goals, and through a list of unsubstantiated promises that guarantee, only in the short term, partial alleviation of obvious and critical community needs, such as basic health and medication.

Local communities are promised, by techniques such as 'educational cartoons' for Filipino peasants, that through 'education' WMC will help local people move beyond their 'caveman' societies into a new realm, surrounded by the technological trappings of advanced industrial societies. Such propaganda materials are simplistic, deceitful and demeaning. On interviewing teaching staff in WMC-funded schools, we learned that once the MoAs were signed, company funding ceased.

The IFFM also discovered that WMC had mimicked the 'community organising' techniques of church and peasant groups, by implanting their own 'community organisers' into these communities to indoctrinate key individuals, with the aim of delivering further agreements. These strategies have created tremendous confusion, violence and alienation within these groups. They are chillingly similar to the tactics of confusion and disempowerment that WMC has used against indigenous peoples in central Australia (see chapter 12). WMC even used the same anthropologists. In 1999 WMC's actions in Australia prompted the Arabunna people (the traditional owners of the land that holds the massive borefields that feed WMC's operations at Roxby Downs) to take out a civil action of genocide against WMC Managing Director Hugh Morgan (see Walker and Crowley 1999).

The second key operational ingredient is cooption and assimilation of all opposition. Big mining companies worldwide are vociferous supporters of the 'wise use movement' (see chapter 11). This 'movement' is a business response strategy to harness opposition. WMC follows the blueprint almost to the point of embarrassing directness. In Australia, as a member of one of WMC's 'community advisory panels', I have experienced first-hand this 'corporate cat herding'. In the Philippines, WMC has created a community monitoring committee (CBEMP), which oversees the environmental monitoring of the company's exploration phase. The committee operates under the leadership of the cousin of Tampakan Mayor Barossa (the Mayor is an outspoken supporter of the company). There are no scientists on the CBEMP. Members are trained briefly at a local high school in basic data collation techniques, and this data is then sent off to the multinational Dole Chemicals for independent analysis. The company 'self-monitors', and has made a mockery of this process, reflecting the obvious lack of respect it has for the local communities.

WMC is diligent in selling the concept that it operates on 'best environmental practice'. Unfortunately, the reality is far removed from the public relations. When it asked whether WMC Philippines would release the results of its own corporate, scientific analysis on its projected options for tailings (mining wastes) management, the IFFM was informed that this information was not 'public' despite WMC's constant claims that it is a 'transparent' organisation. What was ascertained from a meeting with WMC staff at the Tampakan base camp is that the company is seriously considering dumping its tailings directly into the sea. 'Marine disposal', an operation that is illegal in Australia, would bring unknown hazards to the thousands of fisherfolk who pursue a subsistence living along the coastal area. Again, WMC has shown scant respect for the traditions of the Filipino people.

Transnational mining companies such as Western Mining have been given unreasonable advantages in the Third World. In the Philippines, WMC has been granted a five-year tax break, without giving any promises about reinvesting profits in the local economy. With 70 per cent of the population being peasants, none of the profits will be used to alleviate the conditions of advanced poverty that the majority of these people experience. Australian transnationals escape the jurisdiction of nation-states, and deal in a world where cyber-trading and capital flight result in an ever increasing gap between rich and poor. Where once the people of the Philippines fought the dictatorship of Marcos, they now face the often invisible but omnipotent power of transnational capital.

I returned to Australia, promising to spread the findings of the IFFM in the country where WMC has its headquarters and where its shareholders mostly reside.

Why do I tell this story here, at the beginning of a book on the Australian environment movement? My trip to the Philippines allowed me to stand back from the domestic Australian situation and see the movement from the outside. I learned several key lessons, and others were reaffirmed. First, the dominant experience of the environment movement in Australia has been an inward-looking one. Although there are some telling exceptions such as the Rainforest Action Network in Lismore and Friends of the Earth, Australian environmental politics has been dominated by domestic concerns. There has been much talk of global ecology, but there has been insufficient external focus.

The Philippines experience also reaffirmed for me the importance of informal networks. In this book, particularly in part one, I deal with these networks, the real place where social-movement politics takes place. Obviously, these networks must also be outward looking, crossing nation-state boundaries. With the Australian government showing little interest in monitoring the operations of Australian companies overseas, the environment movement must increasingly fill this gap by developing its own transnational networks. Even in my organisation, when I announced my intentions to travel to the Philippines to track WMC's operations there, some members objected: 'But this organisation isn't an international one, it should solely interest itself with issues within South Australia.' I explained that it was a South Australian issue. WMC had its head office in Adelaide and its largest mine in South Australia. Through sharing information with other movements fighting similar battles against this company, our own campaigns would improve.

The environment movement in Australia has come to perceive itself in recent times in the light of a misconstrued notion of postmodern space, where all possible avenues are to be pursued simultaneously. Unfortunately, the strategic weapon of traditional adversarial politics, of disengagement and dissent, has been increasingly forgotten. My trip to Mindanao reinforced the point that 'nice negotiations' with an adversary were not always appropriate. One image stands out in my mind here. I was with a Catholic parish priest just out of Tampakan and we were attempting to cross a river at a time of a flash flood. A WMC four-wheel drive appeared, and an employee offered us a lift through the raging waters. Coming from a society and social movement that were used to dealing with the company, I was about to accept the kind offer, when the priest bluntly answered in the negative for both of us. After waving the vehicle on, we plunged into the murky, fast-flowing waters. On our way back, the priest explained: 'Tim, when battling against an adversary of equal strength, engagement and negotiation is a useful tool. But when the adversary is far more powerful than you, usually to engage is to lose. The only power the people of Mindanao have against WMC is their own solidarity. Once some of us deal, then trust is broken within our own ranks, and we lose our key source of power.'

Togetherness and solidarity are only possible when there is a sense of community. Most Filipino environmental strategies revolve around what is termed 'community organising'. Green activists emerge from a community, as well as being sent 'into' a community to 'empower' local people. The community then educates and organises opposition to the company. Whether these concerns are introduced by environmental non-governmental organisations (NGOs), or whether they evolve from within, both directions of empowerment demand the existence of a definable community. In Australian society, community per se has been eroded.

Another thing I came to realise was that the media in the Philippines were far freer and more critically diverse than our own media. Although representing my state peak umbrella organisation (which is mainstream), I was unable to have any of the IFFM's findings published anywhere in the mainstream press. Our press is extremely sanitised and this reflects our own sanitised society. Australians must recognise that dissent here is almost non-existent, and that any message opposing the dominance of Mammon and its ideology of free markets will rarely be published. Increasingly, green activists are being sued for speaking out against powerful market forces. My own organisation continues to defend defamation suits brought against it by these interests.

The environment movement in Australia has been dominated by issues that do not truly include the human dimension (Doyle and McEachern 1998). Social and environmental justice issues have been rarely considered, and nature continues to be regarded as something separate from people. The Australian movement most closely resembles the North American movements (with some important exceptions), as our whole society increasingly does. What is remarkable about the environment movement in Australia is that it has so far survived the attack upon society, with citizens considered to be consumers, and communities seen as amalgams of separated individuals. To continue to play community politics when the very existence of the political medium—the community—has been almost decimated, is an incredible achievement. The environment movement in Australia remains the most powerful dissenting social movement in our society, continually challenging both politics- and business-as-usual.

NOTES

1   These concepts are developed later in the book. They are sometimes used to describe Australian environmentalism. On occasions they are inappropriate when describing the characteristics of environment movements, or prescribing future strategies.

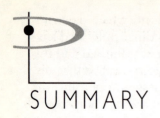

SUMMARY

The environment movement, both in Australia and globally, has impacted upon numerous levels of politics: in transnational business, in parliaments and legislatures, in the bureaucracies of government and inter-governmental departments, in political parties, in legal and other adversarial systems, in the politics of social movements, in non-governmental organisations, in informal groups and in networks of personal relationships.

Apart from being an environmental vanguard, the movement has maintained a radical, oppositional edge, which continues to define modern environmentalism. This is necessary given that many so-called 'green' government and corporate activities have often confused and diffused environmental problem-solving to the point where solutions often become meaningless and ineffectual.

There has already been quite a lot written about the greening of governments and corporations in Australia (Papadakis 1993; Doyle and Kellow 1995; McEachern 1993). Indeed, to focus on these powerful sectors of our society is crucial. Governments and corporations would not be remotely green, however, without the existence of the more non-institutional parts of the environment movement. The environment movement began outside institutions, in the minds and actions of ordinary (and great) people who perceived environmental issues and problems that were largely ignored by mainstream politics.

The movement includes individuals, networks, informal groups, formal organisations, institutions, and corporations. It exists in both the private and public sectors, though this distinction has become so blurred by the lack of separation between the state and large corporations that it has become an almost meaningless form of differentiation.[1] 'Environmentalists' comprise the environment movement. I define them here, in a rather 'emic' fashion, as those persons who refer to themselves as environmentalists.

The movement is present in most parts of our society, both institutional and non-institutional (Diani 1992; Pakulski 1991). Though recognising the institutional elements, this book concentrates on the non-institutional elements of the environment movement. So when referring to 'the environment movement' I am usually commenting on those parts of the social movement that are largely outside institutions and political parties, which do not function within the state and do not work within corporations.[2]

Consequently, this non-institutional focus includes informal networks and groups, and the more formal non-governmental organisations of the movement used as case studies in chapters 5 and 6 of this book. These networks, groups and organisations are usually non-profit oriented. They have fascinating histories and internal political processes, which have not been sufficiently reviewed and analysed. This is quite incredible given that many people consider their actions to have profoundly influenced societal attitudes and politics over the last three decades.

To say that the environment movement has had an impact on nearly every facet of our lives sounds trite, but it is nonetheless true. Any list would be incomplete. Some of the issues it has addressed include: global warming, nuclear disarmament, global security and sustainable peace, pollution, energy consumption, population control, workplace and domestic environments, air and water quality, cultural and biological diversity, gender relations, poverty, social justice, democracy, the rights of indigenous peoples, the rights of non-human nature, the preservation of wilderness, the re-establishment of communities in urban areas, permaculture and other agricultural techniques, alternative lifestyles and non-violence. Stephen Bell (1995, p 30) writes:

> It is difficult to overestimate the gravity of the problems. The Gulf War was essentially a resource war. Land degradation, the destruction of entire river systems, the alteration of global climate systems through the green house effect and ozone depletion, resource scarcity, desertification, the seemingly exponential growth in world population, and habitat and species destruction are all major problems which are rapidly getting worse, not better.

Despite this diversity, this book will concentrate on two broad issues: wilderness and anti-nuclear campaigns. These issues incorporate many, but not all, of the concerns Bell raises here. Whatever standards one uses, these two traditions have been the most dominant over the last 30 years on this country's green agenda.

The book is divided into two parts. Part one, 'The movement', presents conceptual and theoretical tools that can be used to understand the environment movement. I discuss both the informal and the formal realms of environment movement politics. It is important to remember that different political, collective forms—such as the network, the informal group, and the organisation—foster specific types of political ideologies and activities within their parameters and, in turn, these different concepts and structures often shape the level of this activity.

In chapter 1, I present a theory—the 'myth of the common goal'—to explain the existence and activities of the environment movement in Australia. It is also partly applicable to all social movements. This concept challenges traditional political theory while building upon more recent innovations in the social movement literature. It advocates a non-teleological perspective: the environment movement is portrayed as a palimpsest, as a series of amorphous networks with no overriding collection of goals. Rather, it is the different networks that are goal-driven. On occasions, certain networks become so powerful it is assumed that they constitute the entire movement and that their goals are shared by all participants. This has enormous repercussions for the way environmental politics is played and conceptualised.

Participatory democracy in action.

In chapter 2, I describe the politics and the distribution of power in these networks. This chapter also examines the style of politics—and the rationale that supports it—of informal groups that deliberately abstain from institutionalisation. The pressures to formalise are many, but some groups manage to resist these pressures in their pursuit of more radical and democratic goals. I present as examples the Nomadic Action Group (NAG), Earth First! (EF!), the EcoAnarcho-AbsurdistAdelaideCell?! (EAAAC?!) and the Kumarangk Coalition.

Another aspect of informal politics, direct action, is introduced in chapter 3. This is the most overt form of informal politics and networks, groups and organisations often pursue it when they believe that the 'legitimate' pathways defined by the state no longer enable the achievement of environmental goals. The chapter also examines the tensions within the movement between non-violent and more militant direct actions.

Chapter 4 reviews the diverse ideologies of the environment movement. I argue that 'nature' is a largely socially constructed concept. The qualities of nature have been invoked for political purposes throughout history. The environment movement is not new in its attempts to invoke nature in its support for social and political change. What is unusual, however, is the environment movement's explicit invocation of natural processes: most specifically the pursuit of wilderness utopias in which humanity recognises the political rights and values of non-humans. It also includes key social and political concerns that are not readily advertised. These concerns are global in nature, and include participatory and more representative forms of democracy, social equity, non-violence, and the ecology of nature itself. This chapter critically evaluates green theoretical responses from different points of the traditional left–right political spectrum, using the case of green economics as an illustration.

Chapters 5 and 6 focus on the politics of formal environmental NGOs. Some seek a degree of political formalisation without entering the purely institutional political forms of government, political parties, bureaucratic and corporate systems. These participants join the more formal organisations. In most popular perceptions, the Australian environment movement is seen exclusively as a series of green non-governmental organisations. This is largely because pluralist perceptions of politics still dominate our society,[3] and this form of politics is more visible than that which takes place within and between informal networks. The reality, however, is that the formal parts of the movement are just one important form of political activity.

In chapter 5, I discuss key characteristics of all formal organisations, and make a distinction between primary and secondary green organisations. I offer two case studies of international NGOs operating in

Australia: Greenpeace and Friends of the Earth (FoE). Greenpeace has the largest membership of any environmental NGO in Australia, although members are mostly inactive. FoE has a small membership and works in a rather anarchistic fashion, but its members drive the organisation.

Chapter 6 concentrates on two national-level NGOs: the Australian Conservation Foundation and The Wilderness Society. Both have undergone major changes in operation over the past two decades and have experienced several periods of expansion and retraction. There will also be a brief exposition of the Conservation Council of South Australia. In each state of Australia there is a peak environment organisation, which is a broad coalition of formal organisations.[4] These peak organisations—often referred to as conservation or environment councils—provide infrastructure and support to a whole range of other organisations, and make the running on key environmental issues decided by their members. In many ways, these peak organisations are a hybrid political form based on broader social movement models designed to provide solidarity and coordination (while tolerating diversity and ambiguity), as well as operating in the world of formal constitutionalised organisations.

Part two, 'Three periods of Australian environmentalism', comprising chapters 7 to 12, divides the account of the Australian movement since the 1960s into three distinct stages. Each period has two chapters devoted to it. First, there is a description of the dominant political ideology of the state towards the environment in the period. Second, I analyse each period using alternative policy-making models: pluralism/structuralism, corporatism and postmodernism/poststructuralism. Finally, I examine the movement's responses and strategies during these periods.

The discussion does not purport to comprise a history. Instead, the basic chronology differentiating between three specific eras simply provides a primary ordering system. It is not meant to be rigid, but it enables the identification of some broad but clear trends, both in the ways governments and the corporate world have understood and managed environmental concerns, and the manner in which the environment movement has responded.

Table 0.1 illustrates the key dates and characteristics of these three periods. The first period includes the first 20 years, up to the mid-1980s. During most of this period, Australia had conservative federal governments, but the period included the important Whitlam government of 1972–75. By the mid-1980s the 'accord' style of politics, championed by Labor Prime Minister Hawke, began to dominate environmental policy-making. The third period, beginning in the mid-1990s, describes the most recent trends during the re-emergence of conservative politics under Prime Minister Howard.

TABLE 0.1
THREE PERIODS OF ENVIRONMENTALISM IN AUSTRALIA

Period	Date	Dominant ideology	Models	Strategies
1	1960s to mid-1980s	Unrestrained use	Pluralism / structuralism (outsider politics)	Dissent/mass mobilisation and lobbying of govt. Responding to govt
2	mid-1980s to early 1990s	Sustainable and multiple use	Corporatism (insider politics)	Working with govt to formulate and implement env. policy
3	mid-1990s to present	Wise and sequential use	Postmodernism (bypassing the state)	Working directly with or against business and other sectors

Note: This division into periods should not be seen as rigid but as offering a guide to trends.

Two major purposes, therefore, are fulfilled in this story's telling. First, it paints the political context within which the movement is operating. Each period has a clear management ideology that explains the dominant attitude of the state to the environment and its management. These frameworks impact upon different styles of policy-making, with different political outcomes, in real terms.

The ideology of unrestrained use typifies the position of the state in relation to the environment before and during the first period. Indeed, it was this position that saw the emergence of the environment movement as it challenged conventional, destructive attitudes towards the environment. Sustainable and multiple use began to dictate environmental management in the second period. This ideology was developed predominantly by business interests in a bid to incorporate environmental concern into business-as-usual. Still, this era saw the emergence of many cross-sectoral, round-table forums set up by the state that genuinely attempted to provide representative and scientific management of nature and its resources. In the third period, wise and sequential use began to dominate management initiatives. This ideology, derived from the USA, is a strange mixture of conservative morality and radical libertarian economics. This ideology has led to a laissez-faire ecology, with market principles dictating environmental management.

This broader context both reflects and defines movement initiatives. It must be said, however, that the majority of movement initiatives have been reactive over all three periods. The movement has rarely dictated terms, though there have been some telling exceptions. This leads us to the second purpose: to document the ways in which the movement has played its politics given the broader, and always more powerful, political milieus.

The first period saw the movement playing outsider politics. Environmental concern was largely based on direct, oppositional dissent to unrestrained environmental use. Environmentalists, on the whole, were considered deviant 'folk devils', regardless of whether their demands were radical or reform-oriented. Radical environmentalists demanded revolutionary changes, considering the state as incapable of bringing about sufficient social and ecological reforms. These activists concentrated on mass mobilisation techniques and strategies. Their politics and actions can best be understood using structuralist models.

More reformist environmentalists, though still very much on the outer, demanded legislative change to enable the state to manage the environment more effectively. Their major strategy was one of lobbying elites. They succeeded to the extent that an unprecedented range of legislation was passed during these early days, particularly in the 1970s. Pluralist public-policy models are useful in explaining these political initiatives.

The second period begins after the mid-1980s, and can be characterised using corporatist models of power. Although the radical wings of the movement have continued to play oppositional, outsider politics through all three periods, many other parts of the movement began to deal more closely with the state at this stage. Dominant and mainstream green NGOs became incorporated into the Labor government's policy-making processes and agendas.

The third period emerged during the final days of the Keating government, but is more purely characterised by the reign of the conservative coalition government led by Prime Minister John Howard. This period has seen the state reduce its role as environmental legislator, monitor and regulator, and again setting itself up in active opposition to environmental concerns. In reaction, the movement has often been forced to bypass the state and to deal with other sectors more directly. This has produced a challenging time for movement strategists. Postmodern/poststructuralist models of power are sometimes useful in comprehending these disparate strategic responses, while proving problematical when used as prescriptive models for action.

In the conclusion new challenges are posed for the Australian environment movement. The environment movement is at an impasse. I shall argue that the movement must move towards a more human ecological approach while still maintaining an interest in wilderness concerns. Domestic and workplace environments must become part of this new focus, along with shared initiatives with indigenous Australians, who have so often been cruelly left out of the wilderness equation. Poverty and social equity issues must be seen as linked to environmental degradation. This alternative orientation will enable the environment movement to become a broader movement of environmental (human/non-human) change.

I shall propose forms of political resilience within the movement that will allow it to survive the harshest political milieus. The social movement form advocated in the first part of the book is celebrated, as it is capable of maintaining meaningful change that may some day even provide a viable, representative and alternative parliamentary force currently in its infancy in Australian politics. But that is another story.

NOTES

1   It is meaningless in the sense that the public sector was once seen as the exclusive domain of the law, whereas the private was seen as the realm of capital. Though these distinctions are no longer particularly apt, I find that the 'private' and 'public' sectors can still be effectively referred to, although for different reasons. The private sector relates to that part of society that is removed from public scrutiny, whereas the public sector is partly accountable.

2   This line between the institutional and the non-institutional is often blurred, particularly in the case of non-governmental organisations. Their formality (the fact that they operate under a constitution) has led some of them to develop interesting relationships with the state. On occasions they may actually become part of the state.

3   We look later at pluralism as a model of politics. A working definition is that pluralism constructs society as a collection of individuals, with each citizen having equal access to power and resources. The pluralist model often denies inequalities based on class, gender, race and species. Humans are only measured in political terms when they act in a more formal sense (for example, being a 'member' of an NGO or political party).

4   Sometimes non-constitutionalised groupings are admitted into these 'councils', but this is not the norm. When informal groups are allowed to participate, their political rights are normally restricted. For example, they may be given 'associate membership', which allows them to attend meetings but not to vote.

INTRODUCTION

All major wilderness and anti-nuclear issues over the past 30 years have been dealt with by alternatives to the previously established structures of communication and coordination. Certainly, more formal organisations have played a vital role in these issues, but no constitution defines the nature of all their interactions with other organisations and individuals. In the absence of a formalised, hierarchical structure, it has been necessary in each case for the participants to create temporary channels or networks of communication between groups interested in a specific issue.

There is an overemphasis on formal politics in our society, almost to the point where politics of the everyday, informal level ceases to exist. Traditional history has been a story of monumental, 'legitimate' events unfolding within the confines of formal politics. Most people perceive politics to be formal. In this world view politics *is* government, it is that peculiar set of relationships that are forged in Canberra, or in the parliaments of state and territory capitals. Consequently, politics is seen as something tawdry, deceitful, largely a battle between arrogant and egotistical men and, worst of all, a waste of time.

I use the word 'men' here deliberately. Formal political realms are dominated by men, while the more informal political milieus are often inhabited by women. The environment movement in Australia is renowned as a political entity in which women outnumber men. Even in this more 'enlightened' social context, as soon as movement politics starts to take on the traits of formalisation and hierarchy, men seem to emerge from nowhere to assume leadership roles, excluding women from advancement into institutional politics. It is no coincidence that the dominance of women in green politics is one of the factors that explain the movement's continued power and the resilience of some of its more radical, creative and longer-term traditions.

Of course, the politics of our lives can be beautiful as well as tawdry, and all degrees in between. It occurs in our homes, in sporting clubs, in our workplaces, in the streets. Politics simply refers to our relationships to one another, in many different collective and subcultural forms: as individuals, as informal networks and groups, as organisations, including local, national and global ones.

The other infuriating thing about the dominant and severely limiting view of politics is that it portrays the citizen as innately apolitical. In the pluralist system the individual is seen as standing outside the politics of the state. She only becomes involved in 'politics' when there is something wrong: a minor tear has formed in the fabric of democracy. So certain questions are then posed by many traditional political sociologists: 'Why has this person become politicised? What separates this person from the rest?' Such questions look for psychological and socioeconomic factors to explain this mobilisation. In this light, the political actor is not seen as 'the normal citizen': something must have gone wrong for this citizen to become involved in politics.

But if we portray politics as our lives, as citizens who are innately political, then we do not have to look for certain mutant characteristics to justify action. Informal politics is everywhere. It is the driving force of our lives. It provides the spark for change. It expresses our lives. It is here that the environment movement began.

What is even worse in the dominant pluralist view of politics is that it construes informal politics as being somehow illegitimate. Because it is not always governed by constitutions and is not bound by the rule of law, it is seen as peripheral. It cannot be readily controlled, so it is ignored. It is not just governments that attempt to ignore the politics of the informal realm. Non-governmental organisations (the focus of chapters 5 and 6 of this book) dabble in two worlds: the informal and formal dimensions of politics. They are the boundary riders. But even they sometimes forget the importance of informal groups and networks. The Australian Conservation Foundation (ACF) lists hundreds of environmental organisations in its directory, *The Green Book*. Yet it ignores the thousands of informal environmental groups that do not have a constitution. Some groups deliberately abstain from formalising their relationships with a constitution through a fear that this may damage their goals and processes, others merely have not considered it necessary. There are a number of reasons why the ACF does not list these groups. They are not permanent, but appear and disappear as issues do. But aren't our lives also a strange mixture of transience and immutability? If the amorphous dimension is ignored, then what is the chance of change? We seem genuinely afraid of change. We pretend our existence is permanent and gather a material and legal edifice to help us pretend. Our relationships, via institutions, are made permanent. But this is little more than stuffing bones into jellyfish. By denying the politics of informal relationships we deny ourselves, our own existence, and our own death.

An important clarification arises from this discussion. When I use the term 'organisation', I am referring to those collectives that have been formalised with a constitution. Networks, on the other hand, are the most informal, fluid and amorphous collectives. The term 'group' is situated somewhere between these two on the formal–informal continuum, though groups are still clearly informal, as there is no legal framework that clearly defines their operations.

Much of the denial of informal politics occurs because these relationships are difficult to control. And underlying the need for control is a fear of human beings and of nature itself. In this world of fear we need to be controlled, otherwise our 'basic instincts' will get the better of us. Obsession with organisation and institutionalisation gradually removes the human enterprise from humans. We are 'safe' from our own desires and hatreds, but we are also removed from our loves and our dreams. We are effectively disempowered. We hand over our political lives to institutions.

The informal environment movement is much larger than the formal movement. On occasions the former may not wield as much short-term power, but it is more representative of our lives, and it provides the opportunities to work with our creativity and hopes.

Finally, by concentrating only on formal and institutional dimensions, researchers are rewarded with their own respectability. By recognising and studying what others regard as legitimate, they themselves become legitimate. So many government grants are awarded to those who study institutions, in much the same way as the mass media love to publish nice stories about themselves. Someone who writes on the more formal dimensions of politics is then identified as accepting the 'rule of law'. The researcher who studies the informal, the illegitimate, is similarly seen as inheriting the characteristics of her subject of inquiry. Most people do not want the stigma of the illegitimate. This partly explains the lack of study aimed at the non-institutional politics of social movements.

Many analysts have succumbed to the trap of portraying formalised organisations as constituting the most important part of social movements. For example, Geoff Holloway (1986, p 9) argues in his dissertation on The Wilderness Society (TWS) in Australia that organisations constitute the core of the movement, while the rest of the movement 'is some sort of nefarious group that we don't actually meet or have anything to do with unless we want some money'. Pivotal to this view is a conception of the movement as a collection of concentric circles, with the organisations occupying the centre, and all others inhabiting the periphery. This is a rather insubstantial conception of what a social movement is, and illustrates the biases of the organisational sociologist at work.

In chapters 1 and 2, I attempt to provide a more accurate structural depiction of movement politics: a model of the movement referred to as the palimpsest. This model has strengths and weaknesses. It was designed to value and comprehend the input of the more informal, less permanent and, sometimes, more imaginative networks and groups of the movement. Formal organisations are viewed as simply an important component rather than being at the centre of the movement, usually emerging only after the success of certain informal groups and networks.

In this conceptualisation, the movement becomes far larger than traditional analyses would warrant, and there are problems delineating where the study of the movement begins and where it ends. An emic approach leaves the question of the eventual boundaries and membership of the movement to the individual participant. This allows a focus on personal and issue-oriented networks, but presents further analytical dilemmas when discussing 'the movement as a whole'.

For example, acceptance of this description of the movement as diverse and amorphous may lend support to strategists who prescribe overly individualised, neo-liberal responses, denying the necessity, on occasions, of combined action to oppose, en masse, particular attacks on the environment and its people. Description and prescription must be understood here as two distinct purposes of political analysis. Social movements, and societies, are far more than just atomised, individual responses, despite the rhetoric of the market place. Although they encompass a vast diversity of identities and needs, there are times when this diversity must be put aside, even at a symbolic level, as often only through solidarity will political and societal survival be ensured (Axelrod 1984; Ostrom 1990; Ostrom et al. 1999). These questions re-emerge in the final chapters of the book, when I discuss postmodern and poststructural forms of movement politics.

Even within the wilderness and anti-nuclear traditions of the movement there is an enormous number of networks and campaigns. Therefore, any work that strives to understand their politics, their power, will be incomplete and idiosyncratic. Some issues and groups will be selected as case studies, others will be inadequately treated, and some will be unfairly ignored. This problem will be further exacerbated by the book's focus on non-institutional, often informal politics.

These 'everyday' political acts are often not accorded recognition in 'histories', which usually list a chronology of 'monumental' events as defining features of the movement. In contrast, this book aspires to provide a political analysis of parts of the movement, certain strands and networks.

This focus on informal politics requires specific methodologies. This dimension is rarely recorded in mass media, formal organisational minutes and correspondence, and other 'official' depictions of events. To pursue only traditional, positivist research techniques that usually posit the researcher firmly outside the subculture that is the focus of the research is insufficient in this context. To understand the lifeblood of living social movements demands a closer and sometimes total engagement. Many of the issues chosen and the networks that form around them have been selected due to my own involvement as an environmental activist.

This approach provides colour, depth and richness missing from more objective accounts. On the other hand, this approach cannot produce the same rigidity, repetition and cold reason as more positivist, scientific approaches. Simply being part of something is not sufficient. As well as ethnographic, interview and questionnaire-based material, this book is based on official and informal correspondence, documentation and secondary sources that provide extra and more verifiable information. By cross-referencing this material, this study provides a clear picture of the mechanics and organics of the policy-making processes.

This book comes out of the ecopolitics tradition forged at Griffith University in the mid-1980s. At that time, it became clear to a number of us working and studying in the academy that the environment movement in Australia lacked a critical and intellectual edge. Academics and public intellectuals were not sufficiently involving themselves with the politics of environmental activism. In 1986, a group of PhD students, the late Phil Tighe, Ros Taplin and I, attempted with a member of staff, Ken Walker, to build these linkages between the more 'cerebral' and more 'action-based' parts of the movement by establishing the Ecopolitics Conference series. These conferences continue to this day.

This work is also deeply rooted in the field of environmental studies. This broad area of teaching and research is part of a sociopolitical movement. There is a strong notion in environmental studies of advocacy, of a strategic problem-solving focus, of an eclectic methodology in interdisciplinary approaches, of critical and at times radical thought and propositions. Environmental studies has been *for* the environment, in all its diverse and ambiguous orientations. In environmental studies there is a strategic goal: to bridge the gap between theory and practice.

In this vein, ideas inform good actions, and these actions reinform and reshape ideas. Environmental studies, like any field of study, is not politically neutral, but on the whole declares its advocate's role in an overt way.

Due to its interdisciplinary nature, environmental studies demands more than one audience. This book is aimed at introducing some basic concepts necessary for the study of the politics of social movements. It also presents some case-study material that may be of interest to those who already have some competence in the field.

Several other epistemological foundations of this work are listed at the end of the introduction in the form of maxims. Many have emerged from critique of elements of Western thought. For example, such concepts as rational man, teleology, the separation between ideas and actions, universal truths and unidirectional history are contested.

Power is a hotly contested concept in political analysis, and it is also a central one. It is not my purpose to spend too much time debating the intricacies of this concept: this has been done elsewhere (Bachrach and Baratz 1962; Lukes 1974; Smith 1993; Doyle and McEachern 1998). Power relationships are all-pervasive. They occur in families, clubs, schools, hospitals, friendships, parliaments. They may also occur in non-human relationships. Rodney Smith (1993, p 14) writes:

> When people think about relationships within their families, they proba-
> bly think of them in terms of love, or caring, or getting jobs done. This
> is perfectly valid, but it is equally valid to think about these relationships
> as power relationships. To do the latter is simply to take a different per-
> spective—the perspective of the political scientist.

Probably the most commonly understood perception of power in Western, capitalist societies like Australia is individualistic. Those who hold this view understand society as a collection of individual citizens, each wrestling to achieve their will, their interests by using certain power resources, such as economics, knowledge or physical coercion (Smith 1993, pp 16–18).[1] Power, in this manner, has been described by Robert Dahl (1969, p 80) as 'A has power over B to the extent that he can get B to do something that B would not otherwise do'. Power can be enforced through a range of processes: through authority, coercion, force, inducement, persuasion, and manipulation. This theory emerged from North American political science and psychology, and fit in neatly with pluralist, 'democratic' notions of the state.

Critics of this version of power often argue that citizens' lives are usually shaped by external, societal forces. In his widely proclaimed work *Power: A Radical View* (1974), Steven Lukes argues that the plu-ralist version of power is inadequate on three counts. First, there is a systemic bias in any society or subculture due to past actions and the

culturally patterned behaviour of groups, which have often been institutionalised. If power is defined in terms of the capacity of individual
citizens to actualise their wills against resistance, 'collective forces and
social arrangements will be neglected' (Held 1987, p 200). As David
Held argues, it is not unexpected, then, 'that classical pluralists failed to
begin to grasp ... asymmetries of power between classes, race, men and
women, politicians and ordinary citizens'.

Next, power conflicts may not always be visible. Sometimes certain
issues are prevented from appearing on policy agendas due to the traditions of the past or 'the rule of anticipated reactions' (Gaventa
1980). Finally, as citizens are part of collectivities that are subcultures
and societies, these societies have dominant ideologies and mythologies that often prevent certain types of power conflicts from emerging.
Lukes (1974, pp 23, 24) regards this as 'the most effective and insidious use of power'. He writes:

> ... is it not the supreme and most insidious use of power to prevent peo-
> ple, to whatever degree, from having grievances by shaping their percep-
> tions, cognitions and preference in such a way that they accept their role
> in the existing order of things, either because they can see or imagine no
> alternative to it, or because they see it as natural and unchangeable, or
> because they value it as divinely ordained or beneficial?

As well as repressing some ideas and actions, certain societal myths
can champion certain pathways to what is broadly understood as legitimate political responses and endeavours. This portrayal of power is
commonly referred to as the structural view of power. People with differing world views conceive of power differently. In turn, these different conceptions of power inform and promote very different ideas
about the forces that drive or pull individuals, groups and societies.
The pluralist and the structuralist views of power are far from exhaus-
tive. Postmodernist/poststructuralist perceptions of power, for
instance, will be discussed in chapters 11 and 12. All policy-making
models and constructions are to be used as tools for understanding,
not as rigid frameworks within which the 'reality' must be confined.

Models of power often bring with them implicit understandings
about how the world operates. Movement goals and strategies are
often based on these implicit assumptions of what is 'normal political
behaviour'. These assumptions, particularly when widely understood
but rarely questioned, can be usefully referred to as myths. For example, the assumption that the pluralist system of 'democracy' is true and
natural is a very dominant and powerful one, so powerful that people
operating within it experience great difficulties seeing beyond it, let
alone adequately critiquing it.

Attitudes and politics will not change if these assumptions—

myths—remain unquestioned. They constantly legitimate social and political structures and compatible attitudes, values and goals. Myth is the basis for ideology and political organisation. Myths are defined here as a subset of a political entity's ideology. Whether one has a restrictive or inclusive view of ideology also determines how one defines 'myth'. For example, numerous Marxian interpretations of political myth are often restrictive, myth being determined solely by the bourgeoisie and used as a tool with which to repress the proletariat. Inclusive theories of myth are far broader: all people, regardless of class, generate and possess ideas and constructs that include underlying myths.

Although not a Marxist himself, Murray Edelman (1964, p 4) shares this restrictive, structural view of myth. He explains that a myth is 'a collective belief that is built up in response to the wishes of the group rather than an analysis of the basis of the wishes'. Myths, therefore, are based on what people want. Edelman would argue that they inevitably reflect what the ruling class wants, and are diffused throughout the community. He does not suggest that 'elites consciously mould political myths to serve their own ends', but having evolved from a specific system they undoubtedly reinforce what is already established, while denouncing alternatives.

Myths can be both inclusive and restrictive depending on circumstances. All societies and all groupings within societies may have their own myths, however, the myths of the powerful often dominate. Due to the more subjective nature of this analysis, some myths and underlying values that inform the author need to be understood, as they select further questions that frame the parameters of the research.

All questions—particularly those most often and unconsciously asked—shape and confine our understanding of the world. In traditional analysis it would be appropriate to evaluate a movement by measuring its successes against its failures, in a bid to describe its overall 'effect'. Measuring social phenomena in such terms, however, is beset with difficulties. There are so many variables and hidden purposes in the framing of these equations. For example, success in the short term is often at odds with the longer term, lasting change associated with social movements.

Also, success is usually measured by whether goals set at the outset have been achieved. Yet, as any movement activist knows, the best goals are only imagined after the journey has already begun, after we have genuinely entered the field. Too many of the goals and objectives by which we measure 'success' in our society are constructed without any real knowledge of the intricacies of subcultural problems and issues. Aren't many original objectives made meaningless in the unravelling of history? Too often these purposes have been defined outside the real context of people's lives. Also, social movements have many goals. These are multifaceted, often amorphous and full of ambiguities. The traditional success/failure equations are really designed for more formalised organisations and corporations, not social movements.

Despite the fact that life continually offers new paths and new puzzles and pursuits, most formal collective political groups declare their objectives at the outset in their constitutions. Whether or not each collective entity is perceived as successful largely depends on whether it is perceived to have fulfilled these original, static objectives. The motivations behind goal-setting are rarely critically evaluated.

Goal-setting is a learned behaviour. It is dominant in post-industrial, Western societies. It has been promoted to the level of myth:[2] a self-evident truth. Yet it is not necessarily the natural way of human societies. Aristotle is usually credited with having written first about the notion of goal-setting in Western political philosophy. He referred to the rational setting of goals and then the single-minded pursuit of them as teleology.[3] Basically, a citizen operating in the present—point A—projects into the future—point B—trying to imagine what her goals are. These goals are then set at point B, and the citizen then attempts rationally to pursue certain actions that will ensure the type of future—point B—she predetermined.

Teleology is so dominant in our polity that we have great problems imagining how society might operate without it. Teleology is largely socially constructed. It has been very successfully adapted into the mechanics of post-industrialism and latter day capitalism. It is most useful in providing the mythological basis for the pursuit of profitability, centred on axes of time, money, efficiency and effectiveness. Not all human endeavour, however, is goal driven. Social movements, despite popular and many academic perceptions, are not usually driven by common goals. On a personal level, many moments within individual lives are not driven by goals.

Another myth, that of 'rational man', supports that of teleology. Libraries are full of publications about this creature. 'He' is a centrally defining feature of the Western view of humanity and of traditional political science: a discipline born on the coat-tails of a US-led victory in the Second World War. It is argued that humanity is usually instrumentally rational or, at least, should aspire to be.[4] This means that all individuals act rationally on the basis of what is good for them. Even acts that appear on the surface to be selfless (or irrational)—such as helping someone less fortunate—are seen as intrinsically selfish (to make oneself 'feel good'). The rational individual only helps another if there is some associated or greater good for himself, or if there is some pathological reason that explains this 'abnormal' behaviour. Humanity is seen as a set of individuals, each one rationally pursuing his own self-centred goals. It is an atomistic view of society, with each individual competing for the rights to pursue *his* rationally chosen path, *his* objectives.

There are many problems with this vision of political activity. First, it does not allow for the fact that people can and do act for the good of others, with little or no recompense for themselves. Second, large parts of life are irrational or extra-rational, and there is nothing innately wrong with this. There is no essential need to rationalise the non-rational. Some feel a need to control other humans and non-human nature. They fear that without the rational control of all life forces there is danger and chaos. The 'rational man' view of Western society is deeply distrustful of humanity and non-human nature. It is profoundly conservative and neo-Hobbesian (see chapter 4). It sees the *natural* state of nature (which includes humanity here) as nature being at war with itself. The majority of people lead nasty lives that are 'brutish and short' (Hobbes ed. Oakeshott 1962). In order to control and understand, rational *man* continues to seek the 'big picture': hence the grand, unified theories of history. The mind of God is gazed into; the extremities of the cosmos are scanned; the soul is removed from the body, which is then dissected. It is this view of the world that defines success and failure on the basis of a rationally defined future: to select one's objectives and then to pursue them single-mindedly. This is the modernist definition of success: static, short-term and selfish objectives, rationally constructed and removed from context.

True 'successes' regularly have little to do with original objectives. What is more important is that political goals are constantly redefined to suit the circumstances. In the context of the environment movement in Australia, it is obvious that environmental problems can be found everywhere. Never before has the magnitude or number of these problems been so great. To measure the success of the environment movement using traditional, rational-comprehensive tools is therefore frequently inappropriate.[5] Instead, the movement's success is better measured in terms of whether it has managed to remain flexible in the face of amorphous problems that are constantly under redefinition.

Other important questions need to be asked: Has the movement remained an agent of social change? Does it still challenge orthodoxies? Does it truly represent the wishes and dreams of its participants? Is it relevant to social and 'natural' needs? What are its ideologies? What are its mythologies? How is power dispersed, structured and used? How is it organised? What are its strategies? How has it played its politics? How can it best be understood?

Moving on from questions of success or failure, we can say that the environment movement has had an enormous impact on Australian lives. It has challenged the way in which humans see themselves in relation to the rest of nature. It has challenged the ways in which humans deal with each other. And it continues to challenge.

Judgement masquerading as objective analysis is everywhere. Even when we ask simple questions such as 'Is the movement a failure or a success?' we must look at the ideology (defined here as the collection of ideas that are explicitly stated) and the mythology (those ideas that are implicit, self-evident 'truths') that provide the frameworks for understanding. Much of our framing of analysis relies on epistemological judgements.

TABLE 0.2
SOME MYTHS INFORMING THE AUTHOR

1 People are part of nature.

2 Nature/humanity is 'good'.

3 The environment/green symbol is a major one in the late twentieth century. It has come to include issues of social justice and equity, non-violence and peace, participatory and more representative forms of democracy, as well as protecting the ecological wellbeing of both the human and non-human worlds.

4 Ecology is often central to life. It includes issues of indivisibility, irreversibility, urgency, diversity, resilience etc. It is not just concerned with efficiency and effectiveness.

5 Structures and histories often limit the actions of individuals, but sometimes these forces manifest themselves in the form of self-censorship.

6 The type of political structure/organisation within which environmentalists pursue goals will largely determine the types of goals they pursue, and vice versa.

7 History is not uni-dimensional. Our perceptions of what the future will be like dictates the present.

8 The social movement model of political activity is superior in the achievement of longer-term change. Long-term change does not necessitate unified, homogeneous action.

9 Environmental reform and cooption are two distinct phenomena.

10 It is important to act even when there is a real possibility of being partly wrong.

11 There is much to learn from non-human nature. But much non-human nature is a projection of the human onto 'other nature'. 'Nature', as a concept, becomes dangerous when it becomes a justification for human inequality.

12 No one 'model' can provide all the answers.

13 Informal relationships are central to our political lives. Informal politics exists below and beyond formal politics, within and outside the traditional 'nation-state'.

14 Nature/people are not scientific objectifications. People often act beyond rationality.

15 To develop 'resilience' is sometimes more appropriate than to 'resist'.

Certain myths underlie much of the following description and analysis of the environment movement in Australia: these frame the questions I ask. Some of these myths are developed in the chapters that follow, others remain just beneath the surface, and some still remain hidden. Many of these myths will make far more sense after the book has been read, but it is perhaps best to present some of them here.

This book is based on a firm belief in the social movement form of politics and is in part a defence of and a salute to it. The environment movement is capable of continuing to challenge orthodoxies in the long term, in a democratic and useful manner.

It is only since the mid- to late 1980s that the word 'environmentalism', with all its perceptual permutations, has enjoyed any degree of acceptance and respect within institutional politics. Very little opposition existed towards the ideology of 'unrestrained use' before the emergence of the movement in Australia. This ideological opposition did not just emerge because it was essentially 'right', rather, it had to fight to take root within a culture that does not readily tolerate dissent.

I want to turn now to the external mechanisms of power that have silenced movement dissent. As the movement cannot be separated from society, these broader mechanisms have also ensured that there has been inadequate internal critique of the movement's practices. Such a discussion is essential to understanding why there has been such a dearth of critical works on social movements in Australia. As Brian Martin (1992, p 11) puts it: 'Suppression of dissent is a worldwide phenomenon, and is most commonly exercised against political dissidents. Overt suppression is the exception: suppression of dissent works most effectively when it is self-imposed.'

Informed dissent and critique of the unrestrained exploitation and use championed by advanced industrial societies directly contravenes the power of both the state and large corporations. Let us focus for a moment on big business. One of the reasons we do not critically discuss the impact of many corporate activities is that it is a relatively taboo subject. Any criticism is seen as an 'attack on our way of life'. To critique corporations and their advanced industrial enterprises has traditionally been seen as an attack on private ownership of property and its trading of capital assets. Consequently, critique of corporations and their environmentally degrading practices has been simplistically equated with Marxian advocates who are usually seen as 'folk devils'. To be identified with such people is to be alienated, and perceived as an outcast.

This inability openly to critique large corporations is widespread. Even alternative forms of media respond to the coercive powers of large corporations by remaining relatively silent about their environmental malpractices. And what can happen when an individual or group questions the environmental practices of corporations? The case of Alexandra de Blas is most interesting in this context. While completing an honours thesis at the University of Tasmania, de Blas alleged that Mt Lyell Mining and Railway Company had polluted the Queen-King River system and Macquarie Harbour on Tasmania's west coast. Mt Lyell Co. wrote to de Blas's supervisor:

> If you permit any person other than your examiners to read, reproduce or hear the contents of your thesis, in whole or in part, the company may be compelled to take such action against [sic] as it deems necessary to protect its interests pursuant to the Defamation Act 1957. (Montgomery 1994, p 21)

Apart from direct threats made to de Blas if she decided to publish her findings in the broader community, the corporation embarked upon a campaign to discredit both the findings and the methods of de Blas's research. In effect, she was personally attacked. The University of Tasmania's initial decision that the work was defamatory was based on fear of corporate reprisals. The university effectively disowned it, and then reclaimed it, in a series of decisions shaped by lack of confidence. De Blas eventually published her findings in the Australian Centre for Independent Journalism at the University of Technology, Sydney, nearly two years after her research had been completed. De Blas showed great personal courage but, I would argue, not without personal loss.

Most people are reluctant to be identified as folk devils and pariahs. As a result, corporations remain relatively free from societal criticism, while Western governments attract only slightly more. These pressures impact even upon the internal politics of the movement. Movement activists will often censor themselves, using common arguments in justification. I have documented ways in which dissent and discussion of the internal politics of the movement have been suppressed. This process has effectively quashed many opportunities for a better understanding of social movement politics, and the resolution of recurring strategic problems.

Much environmental work is built around the notion of severely limited time. The argument here is based on the assumption that the environmental crisis looms large. With each breath we take, Armageddon draws nearer. A good deal has been written about the impact on decisions in subcultures that perceive themselves to be constantly overshadowed and constrained by time. In a most interesting work concentrating on the decision to launch the ill-fated space shuttle *Challenger* in January 1986, Moorhead et al. (1987, p 542) emphasise the importance of this notion. They write:

Time, as an important element in this model [Groupthink], is relatively straightforward. When a decision must be made within a very short time frame, pressure on members to agree, to avoid time-consuming arguments and reports from outside experts, and to self-censor themselves may increase. These pressures inevitably cause group members to seek agreement.

Most Western cultures are driven to conform to models of behaviour that operate out of a fundamental notion of the scarcity of time. Time is often used as a factor on the efficiency/effectiveness axis that, in turn, is central to profit maximisation. Due to the dominance of such a view, many exciting and long-term initiatives are never even reviewed by mainstream decision-makers. Society remains locked into incremental, short-term activity. Surely, serious environmental problems can only be resolved if we look beyond notions of severely limited time. The crisis factor has replaced the profit factor within many parts of the movement, imposing extreme limitations on social behaviour.

Both TWS and ACF may have some way to go in broadening and improving their decision-making process … It may not be treating humans as well as it should but it sure is getting some whales saved along the way. (Bond quoted in Doyle 1987a)

The notion that there are *legitimate* avenues to question organisational practices within the movement is a furphy. There are few outlets within the movement for the publication of internal critique. Interestingly, when dissent is published in a 'movement journal', it is often viewed with contempt and outrage. Surely valid critique cannot be ignored simply on the basis of denials from those who are, in part, the focus of the critique.

Some believe that the mainstream media image of the social movement is extremely important. Critique of the movement—so this line of thought goes—gives ammunition to reporters to portray it as factionalised and inefficient and therefore not politically viable. There are two fundamental problems with this argument. First, the media seem to at once perpetuate and to reflect dominant social political forms. In a pluralist system the mainstream media are also pluralist in their interpretation of events, and thus fail to accommodate the mechanics and structure of social movements. Social movements are not lobby groups or defined factional interest groups with tangible boundaries. Too much effort is spent trying to construct a hierarchical and unified response. Notions of factions and inner conflict should not be denied, but rather explained as essential elements of living social movements. By maintaining a unified 'media face', movement initiatives are shaped accordingly and are thus distorted and restricted.

Second, if we are continually concerned about 'what the media think', what sells becomes important, not the dissemination of information 'conducive to the wide ranging public debate of policy' (Ward and Cook 1992, p 21). There is often a hidden loss in the human experience when managed media images become all-important. A fear prevails that stultifies and self-censors, an all-pervading concern about what the media will think dominates. It is forgotten that the media are an information tool, not an end in themselves.

To 'question' in our society is seen too often as proof that the individual or group asking the questions is the enemy. It is a world where there are only two sides. So, if you are not one hundred per cent with us, then you are against us:

> Kangaroo took his place before the fire again, but looked aside.
> 'Of course you understand,' he began in a muffled voice, 'that it must be one thing or the other. Either you are with me, and I feel you with me: or you cease to exist for me.' (Lawrence 1989, p 241)

There are often occasions when clear battle lines must be drawn, with problems and forces of opposition precisely outlined (see chapter 12). But central to explaining the widespread perception of the critic as enemy is the fear of internal critique or conflict in our society. Conflict is viewed as unnatural. Paradoxically, conflict is seen to lurk at the very roots of our existence, waiting to strike with evil intent whenever authority is caught napping. Critique, therefore, being based on conflict of sorts, is also seen as bad.

Fear that critique may be used by the 'enemy' is a common theme in the modern environment movement. For example, Greenpeace's reply to a critique of its practices by Hazel Notion was as follows:

> Ultimately, it's disappointing to be attacked on the basis of a lot of rumour-mongering and assertion—those very activities of which the 'forces of darkness' accuse all of the environment movement, and the very activities we should take absolute care to avoid. (Thorstensen and Harris 1992, p 4)

Even when movement members see certain criticisms as legitimate, there are moves to censor them on the grounds that they may be used by the enemy (e.g. 'Arvi Parbo' or 'right-wing think-tanks'). This type of argument is not as severe on the critic as the 'You *are* the enemy' argument, because it gives the critic the benefit of the doubt as to her motives. The critic is now regarded as 'truthful' but misguided in her understanding of the usefulness of the critique to the movement's enemies.

All arguments have elements of the strategic. The critic/activist seeks notional truths, but comes from a position based on value judgements. But strategic concerns do not have to be purely sectarian, nor do they have to dominate. Formal organisations are often scared of critics because the latter diffuse power and underline the disparate character of social movements.

NOTES

1 Rodney Smith in *Politics in Australia* (1993) lists a full range of power resources. These resources are necessary as 'the would be power wielder must possess the appropriate power resources'. In addition, power often generates power resources. Smith lists different types: economic, status, knowledge, solidarity, and those resources based on physical coercion.

2 The use of the term 'myth' is discussed in detail throughout this work. I use it here in the same vein as Murray Edelman in *The Symbolic Uses of Politics* (1964) when he defines myth as something that is widely accepted, and rarely questioned.

3 The discussion of teleology is a central theme developed in chapter 1 relating to the 'myth of the common goal'.

4 I use 'rationality' here in its popular form. There are others. One excellent account of more appropriate forms of rationality appear in John Dryzek's book *Rational Ecology* (1987).

5 For a good account of rational-comprehensive, incremental, and mixed scanning approaches to environmental decision-making, see KJ Walker (ed), *Australian Environmental Policy* (1992), chap. 12.

PART 1

THE
MOVEMENT

SOCIAL MOVEMENTS AND THE MYTH OF THE COMMON GOAL

At midnight the crowd returned to the traveller. 'Time's up', it said, speaking in its 1001 voices, not in perfect unison. 'Answer the riddle. Explain us. What's our definition? What makes us real?' He replied, 'But whatever definition I gave would exclude half of you. It's the lack of definition that unites you.' But the crowd didn't want such paradoxes. It began to quarrel with itself and forgot him. As it receded he could hear shouting and the thump of fists. The air was full of fire. (Rushdie 1988)

Many social movement theories are inadequate to describe the structural 'reality' of the environment movement. These models are premised on assumptions that are questionable, sometimes hidden, and always politically loaded. Many of these models of political activity support the 'myth of the common goal', that is, that movement participation is initiated by individuals with shared, movement-wide goals.

The continuing dominance of the myth of the common goal in social movement analysis produces a number of negative outcomes. It leads to a misunderstanding of the structure of the movement, it fails to provide an adequate framework for understanding the diversity of the movement's ideas and actions, it ignores the importance of informal politics, and it severely limits the potential power of the environment movement in Australia.

SOME TRADITIONAL SOCIAL MOVEMENT THEORIES

Since social movement research became an acceptable focus for academic discourse in the 1960s, there have been many and varied attempts to explain the mobilisation of individuals into the ranks of mass movements. Much of this work has been based on discovering common goals or shared deprivations among environmentalists. In this sense, the theories are strictly teleological. These teleological theories have been dominated, in turn, by two major schools of thought: social psychology and socioeconomics.

The most famous 'collective action' theories have their roots in the sociological theories of the Chicago School (Princen and Finger 1994, p 48). Social psychological theories maintain that social movements attract a particular *type* of person with a particular collection of shared individual goals (Rochford Jr 1982). The range of sociopsychological characteristics reputedly shared by specific movement participants includes alienation (Judah 1974; Seeman 1959, pp 783–91 and 1975, pp 91–122), the search for meaning or purpose (Klapp 1972), the sense of belonging (Mausner 1979), deprivation (Davies 1971), tension (Lofland and Stark 1965, pp 862–74), personal problems (Stark and Bambridge 1980, pp 1376–95), deviance, disorganisation, and the desire to help others (Sills 1971, p 823).

Each theory attempts to explain motivational characteristics—the common goals that are shared by those individuals driven to join social movements. There are weaknesses in these theories. First, most of them make the assumption that social movement involvement is not the norm. Terms such as deviance, disorganisation and deprivation are common in these works. These theories try to uncover the source of mutant behaviour. Yet involvement in social movements is typical of all society. Second, many people who have one or more of the above attributes resist movement participation. Finally, the great diversity of personality-based motivational factors in the environment movement makes attempts to find one psychological factor common to all participants look rather absurd.

Another very popular theory in academic circles relates to the middle-class origins of environmentalists. Alain Touraine's work on social movement theory is widely respected and partly indicative of this stance. He writes that 'The social movement is the organised collective behaviour of a class actor struggling against his class adversary for social control of historicity in a concrete community' (Touraine 1980, p 77). Class-based theorists argue that environmentalists share certain middle-class goals that incite them to political activism. Harry, Gale and Hendee (1969, pp 246–54) write in the context of the USA:

> Membership in the conservation movement appears to be composed largely of the upper-middle class occupations, especially professional occupations. In addition, it is primarily an urban-based movement that is somewhat isolated ideologically from the main-streams of both liberal and conservative political thought.

Research on the Australian experience has offered similar theories. In his study of the movement in Tasmania, Geoff Holloway (1990, p 9) often refers to the fact that environmentalists come from the same social stratum, 'which is characterised by particular age-groups, high-educational achievement, professional occupations, high social mobility and concern for the realisation of non-economic values'. The membership of organisations such as the Australian Conservation Foundation (ACF) and The Wilderness Society (TWS) are used to illustrate this point. The ACF *Newsletter* of March 1984 listed 51 per cent of its members as professional, while a massive 76 per cent have taken part in tertiary education. The 1985 TWS Planning Survey stated that approximately 60 per cent of its members held professional employment, while 50 per cent have had tertiary education (Holloway 1986).

Holloway argues that 'this stratum is well-equipped to participate actively in politics', but he does not explain why the initial contact is made. Using the same premise as Holloway, in an analysis of the conservation movement in England, Cotgrove and Duff (1980, pp 333–51) attempt to uncover a reason for mobilisation. They conclude that a remarkably high proportion of environmentalists who were sampled occupied roles in the non-productive service sector: 'doctors, social workers, teachers, and the creative arts'. On the basis of these results, they contend that not only are environmentalists drawn from the middle class, but from a 'special fraction' of this class 'whose interests and values diverge markedly from other groups in industrial societies'. It is the political and societal estrangement of environmentalists, as a special class, that projects them into the arena of political activism. They argue:

> Hence, their concern to win greater participation and influence and thus to strengthen the political role of their members. It is a protest against alienation from the processes of decision making, and the depoliticization of issues through the usurpation of policy decisions by experts, operating within the dominant economic values.

These class-based theories are simply inadequate for the environment movement in Australia. As I argue in chapter 2, the diversity of individuals and ideologies seems to go far beyond the boundaries of any specific class or interest group. The major flaw with these theories is their limited definition of the environment movement as a collection of formal organisations. The majority of formal organisations operating within the movement may, indeed, be made up of predominantly middle-class members, but so too are most formal organisations in all political spheres.

Traditional social movement theories, however reworked, are simply inappropriate to social movements. Social movements are not large organisations. Interest group theories, based on the study of formal organisations, are wedded to a pluralist view of politics. Pluralism portrays the environment movement as a series of pressure groups temporarily rising to pursue uniform and shared interests until such time as the state satisfactorily resolves the problem that led to their emergence.

The environment movement's ideological heterogeneity also contrasts dramatically with parts of modern political economy, which often espouses the view that successful collective human action must be based on total homogeneity of ideals and goals. Mancur Olson's (1965, p 33) work exemplifies these dominant theories. He writes:

> The achievement of any common goal or the satisfaction of any common interest means that a public or collective good has been provided for that group. The very fact that a goal or purpose is common to a group means that no one in the group is excluded from the benefit or satisfaction brought about by its achievement ... It is the essence of an organisation that it provides an inseparable, generalised benefit.

It is often assumed that if total goal agreement is not obtained, then the human cluster concerned will inevitably dissolve.

These ideas are present in our politics. It is often said that a political party that incorporates more than one 'party line' is self-destructive and doomed to collective failure. Inner conflict, it is argued, is an 'evil' time- and resource-consuming process that makes success more difficult. Tied up in this notion is the concept of utility maximisation (Frohlich and Oppenheimer 1978). Optimal efficiency can only be achieved if goals based on one commonly held set of beliefs—usually that of the dominant regime—are pursued. This accounts in part for the lack of tolerance of dissent within the movement.

There may be another reason for the continued popularity of assumptions about the common origins of movement participants. It may be in the interests of certain antagonists to portray environmentalism as a middle-class movement. Thus environmental concerns are represented as a luxury the working class cannot afford. Those in lower income brackets may come to hate what the movement stands for.

Although class-based theories may illuminate some motivations common to activists in formal organisations—a small proportion of the movement—they are largely ineffective when applied to a broader definition of the movement that includes all levels of formal and informal politics. The psychosocial and class-based theories share a teleological vision, that is, they present collective political activity as inevitably emerging on the basis of shared, rational political goals. Yet the environment movement in Australia has no common goal.

Traditional sociological and political research has transferred concepts based on hegemonic organisational structures to the new analytical arena of social movements. Transported along with these theories are social myths evolving out of, and inevitably protecting, the established order. The myth of the common goal is one of these.

THE MYTH OF THE COMMON GOAL

The 'myth of the common goal' is a *political* myth in every sense of the term. Henry Tudor (1972, p 139) describes its role as follows:

> A political myth may explain how the group came into existence and what its objectives are; it may explain what constitutes membership of the group and why the group finds itself in its present predicament, and, as often as not, it identifies the enemy of the group and promises eventual victory.

The myth of the common goal meets all of Tudor's characteristics. In an extensive series of formal and informal interviews I conducted between 1985 and 1995, Australian environmentalists were asked six questions.[1] The first was whether or not the respondents believed there was something that bound movement participants together and, in turn, separated the movement from the rest of the community. Eighty-two per cent answered that they believed there was something, most thought this was stating the obvious. Also, most respondents, unprompted, replied that the binding factor was 'a common goal'. For the benefit of those who had not voluntarily named the binding factor, the second question was whether this common trait was a broad belief system or common goal. Seventy-five per cent said yes, referring directly to a 'common goal', while others viewed it as 'a kind of collective consensus' or ' shared underlying belief'.

When asked to define the so-called 'common goal', people gave greatly varying responses. Of the 450 participants interviewed, approximately 25 per cent thought that the common goal had 'something to do with conservation or the environment'. Others believed that the movement was primarily concerned with vast social change. Some participants specified that the eradication of an ecologically 'corrupt' lifestyle was the most fundamental, unifying goal. Several participants merely repeated, in parrot fashion, the primary constitutional goal of an organisation to which they were affiliated. A few participants had problems naming the goal they were so adamant did exist. Each participant firmly believed that the particular goal she had identified was *the* factor that unified the environment movement in Australia.

Another interesting result from this questionnaire was that different environmentalists had markedly different perceptions of which groups and individuals constituted the movement. For example, in the initial pilot survey of 125 Queensland environmentalists, 55 participants included the Brisbane branch of Animal Liberation, while 47 denied this group membership in the movement. Similar results were found when environmentalists were asked to comment on organisations and individuals campaigning for nuclear disarmament. Most confusion occurred among participants when they were asked whether or not they thought the Queensland Department of Primary Industry's Soil Conservation Authority was part of the movement. The 'affirmative' and 'negative' votes were equally divided, while the majority could not decide.

The conviction that common goals—based on a collective system of beliefs—bind the environment movement together is not unique to eco-activists. It is a myth with a broad, well-entrenched social base. Many academics writing on the social movement phenomenon constantly reiterate the existence of 'common goals'. Brian Martin (1984, p 110), for example, writes:

> If social action on an environmental issue is sufficiently coherent, organised and sustained, it is appropriate to speak of an environmental movement. In a movement the otherwise 'ad hoc' activities are tied together by a general goal and usually a sense of purpose and unity.

What then is the common goal of the environment movement? Perhaps the objective is unconscious, intangible? Some writers, such as Paolo Melucci (1984, pp 819–35), argue that all contemporary mass movements have the same unifying, primary goal: the new organisational form of the movement. He writes:

> The new organisational form of contemporary movements is not just 'instrumental' for their goals. It is a goal in itself. Since the action is focused on cultural codes, the form of the movement is a message, a symbolic challenge to the dominant patterns. Short-term and reversible commitment, multiple leadership open to challenge, temporary and ad hoc organisational structures, are the basis for the internal collective identity, but also for a symbolic confrontation with the system.

Certain sections of the movement do operate as Melucci describes and these types of operation may be symbolic in that they attempt to redefine the meaning of social action. But other sections of the movement function in a formalised, structured fashion, openly endorsing the existing status quo.

In the absence of an easily defined common goal, new concepts have emerged under new names, but all still assert the existence of common ideological factors that drive the movement's extremities inwards until all is one. One such concept is Murray Edelman's (1964, pp 154–6) idea that there exists 'a common latent attribute underlying a number of values with differing content'. Edelman argues that human goals and ideals can often be portrayed on what he terms a 'Guttman scale'. This scale is defined as an underlying dimension upon which sets of common political goals are based. He gives several examples of these so-called 'Guttman scales'. One such set of objectives revolves around what he terms 'a social welfare structure', evident in the USA when he was writing. He contends that 'a "social welfare" value structure includes items on aid to education, medical care, employment guarantees, FEPC and Negro housing, and public versus private production of electricity and housing' (Edelman 1964, pp 154–6).

The existence of an underlying dimension is at first an attractive explanation in the absence of a definite common goal. It is correct in the sense that many environmentalists are active in many other related fields of community endeavour. The environment movement is, however, far too diverse in its values, political philosophies and goals to suggest that these can be accommodated within a single paradigm based on social welfare or conservation issues. Many individuals who are part of the movement could identify with a number of different value structures.

Piven and Cloward (1979, p 5) use another approach, which, at first glance, appears to support my contention that there are no overriding common goals that unite the movement. They argue:

> Thus formalised organisations do put forward articulated and agreed-upon social change goals … but such goals may not be apparent in mass uprisings (although others, including ourselves as observers and analysts, may well impute goals to uprisings). Furthermore our emphasis is on collective defiance as a key and distinguishing feature of a protest movement, but defiance tends to be omitted or understated in standard definitions simply because defiance does not usually characterise the activities of formal organisations that rise on the crest of protest movements.

Piven and Cloward's assumption that mass movements evolve and, in turn, are held together by a collective defiance deserves attention, but these ideas do not explain the environment movement either in Australia or elsewhere. Of course, collections of various networks, groups, and individuals in the environment movement would share a certain defiance towards the present political system. Many others, however, would not. Piven and Cloward's research concentrates on poor people's protest movements, and although some sections of the movement in Australia share similar class backgrounds to these particular subjects, the movement cannot be delimited on a class basis. Instead, the movement is representative of all social classes in Australia.

Although Piven and Cloward dismiss the concept of a monolithic goal, in a negative sense they attempt to provide one. Their definition of social movements is based on the often quoted assumption that 'it is easier to tell what they are "against" than what they are "for"' (Wilkinson 1971, pp 118–19). But this is not a sufficient base for any conclusion.

THE MYTH OF THE COMMON GOAL AND ITS EFFECTS UPON THE MOVEMENT

The evolution of the myth of the common goal has had profound effects on the environment movement. It has alienated the environment movement from the 'rest of the community'. Despite the fact that the environment movement is representative of all sectors of society, this myth can engender strong opposition to the movement, for it suggests that while the movement's goals bind it, the movement, together, they differentiate it from the rest of society. Particularly during the 1960s and 1970s, the movement was branded as abnormal, subversive. In consequence, society misconstrued the very nature of the movement and perceived it as a threatening external force.

An excellent illustration of this occurred in the Queensland environment movement's 'successful' attempts to maintain Lindeman Island's national park status, when this was under threat from the ruling National Party and East-West Airlines during the last years of Bjelke-Petersen's premiership (Doyle 1987a). In this case the media did not portray the environment movement as politically removed from society, but as representative of it. In fact, during this campaign certain branches of, and individuals affiliated with, the National Party came to share certain goals with the more conservative members of the movement, and matched their concern with collective activity. Thus several National Party branches became part of the environment movement despite the fact that the party played a central role in the dominant development regime.

The myth of the common goal weakens the political power of the environment movement. As we have seen, participants in the movement genuinely believe in the existence of common goals. Also, as each environmentalist has a different perception of the unifying goal/s, they also thus exclude certain sections of the community from the movement. Apart from denying the movement increased potential power, this belief confuses members about the internal mechanisms of the movement, with a matching loss in effectiveness. Most importantly, demands for sameness and goal uniformity are a means of excluding more radical voices.

One good example of this point is the wet tropical forests campaign. In the mid-1980s, the then federal Minister for Arts, Heritage and the Environment, Barry Cohen, was receiving a multitude of separate submissions from large numbers of individuals and organisations within the environment movement. This created 'voluminous paperwork' for the bureaucracy to deal with. A dissatisfied Cohen told environmentalists involved in the campaign to form a single voice, and only then would submissions be seriously considered.

This request is a common one. The government, accepting the myth of essential goal-uniformity, believed that if it demanded to deal with one representative, then this would inevitably play off various factions of the movement against each other, thus sapping it of political strength. Had the movement been conscious of its divergent forms, it would not have consented to this request, but explained that no single group, organisation or individual could possibly represent all those involved in the movement. But because the movement was also caught up in this myth, the following occurred:

- The movement wasted time in attempting to establish a leading group that was truly representative of all groups concerned (this was an impossibility, and the attempt resulted in 'unnecessary' conflict).
- It lost its ascendancy over the bureaucrats by agreeing to conform to their structural guidelines.
- It allowed the government to set the agenda for continued correspondence and behaviour, again taking away some of the initial impetus of the movement's approaches.
- Sections of the movement became alienated, so that many constructive relationships remained untapped.
- It perpetuated the myth, and further removed itself from society as a whole.

Once belief in the myth of the common goal has limited the boundaries of the movement, it then demands that some broad compromise be made between the remaining participants. This compromise inevitably takes the form of some incremental goal or ideology. The revolutionary ideologies of the movement are termed 'extremist' and 'not representative of the broader movement'. The power the movement once had to force dramatic change is limited.

Belief in the myth of the common goal often gives power to reformists who operate within formal organisations and institutions. These organisational activists are also often professionals. In the wet tropical forests campaign, for example (see chapter 10), a small band of professional, organisational activists banded together to dominate many conservation initiatives. As such elites increase their hold on movement politics, representativeness and equality in decision-making diminish.

It is remarkable that the impact of this myth on the movement is still as powerful some 15 years after the wet tropical forests campaign. In mid-1999, during the movement's negotiations concerning the proposed Environmental Protection and Biodiversity Conservation (EPBC) Bill, almost exactly the same scenario unfurled, with the same repercussions: only this time, a different government was in power. The Howard government has been the worst in relation to the environment since the birth of the modern movement in the late 1960s. Senator Robert Hill, a consummate professional politician and diplomat, was handed the environment portfolio to 'keep a lid' on environmental affairs, to remove the environment from its high ranking on the national political agenda. His negotiations on greenhouse issues in Kyoto, on Jabiluka in Paris, and on wetlands in Costa Rica illustrated that he would make an excellent foreign minister, but his silky dealings were weighted heavily against environmental interests. Despite his frightful environmental track record, Hill managed to engage the movement in negotiations over the EPBC Bill, and to create a poisonous rift between major green NGOs. Continuing belief in the myth of the common goal also contributed to producing the rift.

On the whole, the EPBC is an appalling piece of legislation. It effectively guts all the major environmental legislation developed over the past 20 years. Key flaws in the new system include:

- a completely inadequate list of matters of national environmental significance currently included as environmental impact assessment (EIA) triggers such as those relating to climate change, native vegetation, land degradation, water allocation and forestry operations;
- the failure to retain Commonwealth funding as one of the new triggers; and
- the failure to include a ministerial reserve power in relation to EIA triggers (adapted from Environmental Defenders Office NSW 1999).

The Commonwealth has basically abdicated many of the powers it once used to protect important environments, and handed these over to the states, which have always been among the most keen on development. The Environmental Defenders Office (NSW) made an independent analysis of the legislation after its enactment. In its final form the legislation included 500-odd amendments the Democrats asked for in their new incarnation as 'deal maker with the stars', which were 'tabled one day and passed the next, under a special Democrat-government agreement to curtail debate' (Connor 1999, p 9). The Environmental Defenders Office NSW (1999, p 7) writes:

It is worth commenting specifically on bilateral agreements ... In our view, the most significant flaw in the Act is that it permits the Commonwealth to delegate its EIA powers to the States ... It is extremely disappointing that the Government has not chosen this seminal piece of legislation to make a strong, unambiguous statement about Commonwealth leadership in the environmental field. This Act sends mixed messages—and at a time when the extent of Australia's environmental problems is only too obvious, this is something our environment can ill afford.

The legislation was deliberately created as the bearer of 'mixed messages'. Although it excluded from its protective embrace such fundamentals as forests, air, water, and land (is there anything else apart from fire!), it did include some very good provisions for the promotion of biodiversity and threatened species protection. These projects are championed by the more traditional 'nature conservationists', who do not have broad social and environmental goals but can pursue 'triage' for threatened species and play around in farmers' paddocks planting trees indiscriminately, all the while muttering the mantra of biodiversity. All these things can be achieved through politics- and business-as-usual, and Robert Hill knew this.

The movement took a 'suck it and see' position on the legislation in the lead-up to the final vote in the Senate. Professional officers of most of the mainstream environment organisations had met twice before these negotiations at Peter Garrett's house at Mittagong, and had decided, without sufficient consultation with their elected officials, let alone their members, that legislation related to the goods and services tax (GST) and the EPBC Bill were the most important items on all major environmental organisations' agendas. Giving prominence to the GST and EPBC issues, already strongly differentiated on party-political grounds, would relegate all other issues to the bottom of organisational agendas.

There were many positive things about these team-building exercises, but a crucial lesson still had not been learned: the professional employees of mainstream organisations are not the movement, but simply a very important, committed and significant part of it. At these meetings it was agreed that a 'movement-wide' negotiating team be selected to deal with the government. Both the GST and EPBC negotiations saw a new kind of democracy emerge among mainstream green organisations: democracy by e-mail. It was decreed: 'If you don't reply, then you must agree with us, and you've only got until this afternoon to make a decision.' This was how the movement would reach its common goal.

One development that we saw in this instance was the opening of a fundamental gulf in goals between the old-guard 'conservationists' and the 'environmentalists'. The goals of the two groups were different, as were the political processes pursued to achieve them, yet they were still part of the same movement.

The yearning for a movement-wide, common-goal position by a negotiating team, which was not elected by the movement, coupled with the selection of the GST and EPBC as dominant agenda items, led to the movement's capture by party politics, yet again. In a letter to Queensland Conservation Council (QCC) member groups, Imogen Zethoven (1999), QCC coordinator explained away insufficient consultation with her member groups as follows:

> A number of QCC member groups have expressed concern that neither they nor the QCC Executive were informed of the negotiations. Due to the lightning speed by which the Queensland Conservation Council coordinator became involved in the negotiations at the eleventh hour and due to the confidential nature of the discussion with the political parties, this simply was not possible.

All political parties wanted the movement's support. The Labor Party moved against the EPBC legislation, along with the Greens. The Liberal, National and Democrat parties supported the legislation. All mainstream organisations, including my own, CCSA, were asked to take a 'position'. The ACF and TWS sided with Labor and the Greens, while the World Wide Fund for Nature (WWF), International Humane Society (HSI), Queensland Conservation Council (QCC) and Tasmanian Conservation Trust (TCT) sided with the Democrat–conservative coalition. Inevitably, most green organisations were placed in a lose-lose situation. It was particularly difficult for state-based umbrella organisations, which are broad amalgams of

Forests have dominated wilderness issues in Australia for a generation.

informal groups and formal organisations that span the political spectrum. If they endorsed a party position, they lost support from many of their member groups and from the 'other party' past the next election; if they remained non-committal, they were side-stepped.

Cam Walker, national liaison officer for FoE Australia, wrote in frustration to the executive officer of QCC, Anthony Esposito, expressing his concerns about the outcomes that QCC and others had so hastily pursued:

> It is FoEA's opinion that the actions of QCC, WWF, HSI and TCT have furthered the agendas of big business and Senator Hill, who wish to marginalise community-based conservation groups by splitting the NGO community and by establishing negotiations with select individuals from within 'the 'big end' of the environment movement. (Walker 1999, p 2)

Finally, in the search for a unifying goal, public criticism of the position of the QCC, WWF, HSI and TCT was quietened, as professional environmentalists maintained that 'we are all after the same thing anyway'. When mild criticism is raised, the powerful elements of the movement now dealing with Hill (which will lose power when Labor gets in) threaten to split, thus weakening the movement and sapping its most important resource: its solidarity, its diversity and its numerical strength. Steve Baker (1999), one of the few independent green lobbyists in Canberra writes:

> WWF and HSI are international groups which work closely with governments and industry. They have high public profiles, and are considered to be 'realistic' and work with organisations like the IUCN and the World Bank. As I understand it WWF has basically said it intends to be part of the EPBC bilaterals (BLs) regardless of whether it gets movement support or not, even suggesting that the movement should split. I do not believe that WWF should be allowed to participate in the BLs. They showed with their EPBC orchestrations with the Democrats and the Coalition that they have little regard for democratic process within the movement and I would expect no less when it comes to negotiating the BLs.

The myth of the common goal fulfils Tudor's criteria of political myth. It is a restrictive myth as it often serves the interests of the powerful. It was born out of pluralist society and, more often than not, supports it. It provides a scenario that almost demands that elites take control. This is what happened in the wet tropics and EPBC cases. Cuthbertson (1975, p 174) writes of myths as necessarily serving the status quo, primarily through the processes of legitimisation:

> The political myths relieve the uncertainties and insecurities of power-holding. Myths erase the bar sinister from the usurper's armorial bearings.

They pardon the crimes behind the crowns. Myths minimise the amount of 'naked power' required to retain control of government. They ordain, validate, and substantiate the 'powers that be'.

THE BONDING FACTOR OF THE MOVEMENT: UNDERSTANDING NETWORKS

Escaping from the confines of the myth of the common goal allows us to view the movement in a far more realistic way. Figure 1.1 represents numerous informal groups and formal organisations involved in the wet tropical forests campaign in Queensland. This figure does not list all participants in the campaign, which would be impossible. Rather, it provides a conceptual aid to the discussion. The environment movement in Australia is a web of networks. Each network may include individuals operating in groups and formal organisations, as well as other networks. Within the movement there is no unifying ideological concept, no goal that binds it together. On the contrary, each entity has distinct objectives, arising from a vast array of differing philosophies of life. These separate entities, however, quite often share similar goals with one or more other groupings and individuals within the movement. It is this characteristic, matched with collective activity and identity, that constitutes the movement. Collectives and individuals with similar goals are represented in the diagram as being close together. It is quite possible for two separate entities without common goal characteristics to share objectives with a third party, and thus be indirectly be linked in the movement. Individuals or entities connected to each other in this fashion are situated upon what is termed a 'goal chain' (see Figure 1.2). This particular goal chain uses the 'left–right' political spectrum. Situated on the far left of the chain are those whose primary concern is non-violent action. The group NVA also espouses 'consensus decision-making as central to its anti-hierarchical critique of society' (Holloway 1986, p 6). NVA has firm roots within the environment movement, particularly The Wilderness Society, and shares obvious goals with the anti-nuclear lobby.

Within the same movement is the organisation Campaign For Nuclear Disarmament (CFND). While arguing against the continued mining and export of uranium, this group of people also promotes a 'soft energy path'. This entails 'small-scale renewable technologies' and 'local self-reliance and participative decision-making' (Martin 1984, p 112). This organisation actively campaigned against the National Party government in Queensland, which retained power through the use of the infamous gerrymander, which denied individuals equal social and electoral rights.

CFND shares obvious common goals with those espousing non-violent action. CFND was also a member of the Queensland

Figure 1.1

Web of some groups and organisations involved in the wet tropics campaign

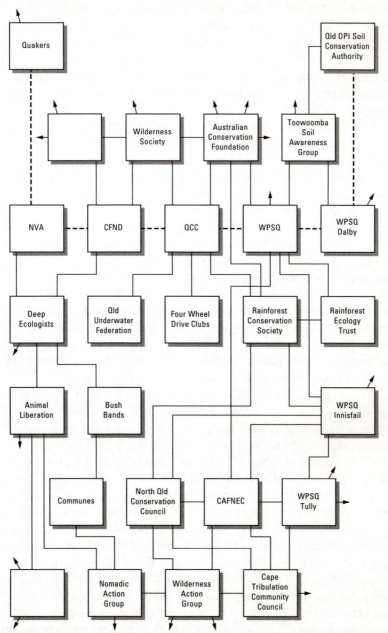

Note: Broken line indicates goal chain. Groups and individuals in the centre of the diagram are no more important in the movement than those represented on the periphery. This diagram concentrates on demonstrating some goal similarities between human clusters. Individuals are also part of these networks, but it is difficult to represent them. All links are based on the author's knowledge of goal similarities. Compiled by the author.

Conservation Council (QCC). This political association benefits the two organisations in different ways. QCC provided CFND with contact to a multitude of other formal organisations, and made available the facilities of an environment centre. On the other hand, the link with CFND created the impression that QCC was indeed representative of a wider spectrum of groupings, which, in turn, made possible the centralisation of real resources and power. As well as having divergent interests, both organisations shared a common goal: a nuclear-free society.

The Wildlife Preservation Society of Queensland (WPSQ) is next on this particular goal chain. WPSQ is also a member of the QCC. The two organisations share similar objectives. Both are central administration bodies. WPSQ has 20 branches in Queensland, each of which, though distinct in their evolution and make-up, share some common objectives with their central administration.

WPSQ Dalby is one such branch. Its primary concerns are soil conservation and other measures that will ensure continued profitable agricultural activity. The majority of its members were rural-based supporters of the existing National Party government, with strong feelings in favour of, for example, small government, the family, uranium mining, and the supremacy of *man* over *his* environment. On the extreme right of this goal chain is the Queensland Department of Primary Industry's Soil Conservation Authority. Despite being a governmental organisation, the authority is part of the same goal-oriented soil conservation network that includes the Dalby branch of WPSQ.

Within this goal chain are organisations—and more importantly individuals—that are philosophically and politically diametrically opposed. Such organisations do not share common goals based on a similar belief system. To give one example, the relationship between WPSQ Dalby and CFND is not a direct one. WPSQ Dalby is in fact so opposed to nuclear disarmament that it refuses to join the QCC largely because of QCC's disarmament policies. Despite their lack of goal uniformity, these two organisations, among many other ideologically diverse individuals and organisations, are part of the same movement.[2]

These indirect links, through a goal-oriented relationship with a third party, are more than just perceptual. They are extremely mechanistic. They represent a bonding factor even in the absence of an overriding common goal. These links provide communication between all members of the movement. This communication is not always intentional, but it does have an effect. For example, whether these organisations acknowledge it or not, WPSQ Dalby receives information originating from CFND—and vice versa—even if it has been diluted as it has travelled across the goal chain. As well as correspondence there is information exchange in individual and group discussion.

Figure 1.2
Goal chain

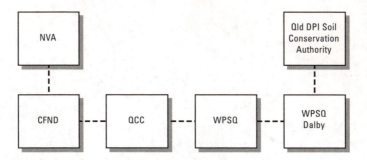

The goal chain shows how organisations that would normally claim to be totally separate, if not opposed, can be linked through their association with other groups. Compiled by the author.

tropics, but around all the issues pursued by movement participants. The palimpsest includes formal organisations, informal groups and individuals. It is a snap-shot in time of a social movement that is alive, always moving, always redefining its shape and force. It is a three-dimensional space, although the added dimension of depth is not apparent here. With the addition of depth the conceptualisation is even more complicated, as each individual is usually a participant in a number of networks simultaneously. Some networks may appear instantaneously, and then disappear almost immediately, others have far greater longevity, existing for the length of an entire campaign or longer.

The environment movement in Australia, then, is fundamentally made up of a plethora of informal networks that link the organisations, groups and individuals. Although the *movement* does not have a common goal, each *network* does. The intermeshing of these networks makes up the movement. Doyle and Kellow (1995, p 90) describe this phenomenon:

> Each network is different, as its definition relies on the perceptions, biases and power plays of the initiator(s). The key differential variable of the network, however, is the common goal or ideology that binds the participants together. In most cases these goals or shared values are specifically issue-oriented, whether based on an environmental campaign or a particular type of political system. These interconnecting lines ignore organisational and group boundaries, for these networks fundamentally are concerned with relationships between individuals operating inside and outside other formal and informal collective forms.

Figure 1.3
The palimpsest model of the environment movement

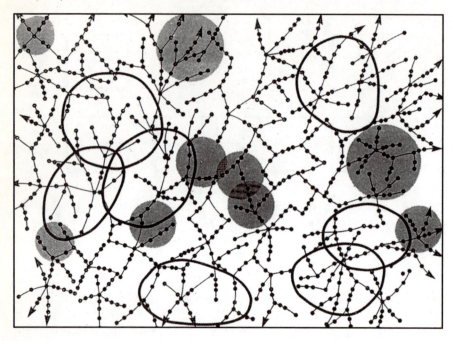

KEY

- • Individual
- ✳ Network
- ● Group
- ○ Organisation
- ▭ Movement

Compiled by the author.

Networks are informal collective forms, they connect individuals. Formal organisations and other collective forms may be included in any network, but their inclusion still rests with individuals. Obviously the prior positioning of the organisation in relation to the network is an important factor in network formation. If the organisation becomes part of a network it is because one or more of its members are part of it, not the formal organisation as a separate political entity.

There is now a recognised split between what has become known as the 'American' and 'European' approaches to social movement theory (Morris and Herring 1987; Klandermans et al. 1989; Neidhardt and Rucht 1990). Although there are important exceptions in both categories, this division can be usefully employed here. The more traditional approaches were dominated by the American theories of collective behaviour, resource mobilisation, and those relating to political process (Diani 1992, pp 3–4). Although there are important and fundamental differences between all of these schools of thought, many of their advocates do insist that a shared teleological dimension, with other factors, ties participants together, defining the movement. The more European tradition, which is often referred to as the New Social Movement approach, does not place such an emphasis on commonality of purpose, but rather on the importance of networks as *the* defining factor of this modern social phenomenon. Paolo Donati (1984, pp 837–59) writes of this phenomenon in the context of Italian mass movements:

> Collective identity can only be formed through concrete and significant interaction between individuals. Its bargaining and its formation are carried out through pre-existing ties and networks, through everyday relationships and collective identities which are always present in the social system and which come to be changed and reshaped through the bargaining process.

These more recent, European (and often Italian: Melucci, Donati, Della Porta, and Diani) theories are supportive of my contention that the movement does not have uniform goals, but is rather a collection of diverse, informal networks. These theories argue that the existence of these networks is a necessary precondition to the formation of symbolic identity, rather than the networks coalescing around a pre-existing series of rational goals. Most people join a movement network, therefore, because they already know someone who is active in it. In my survey of environment movement participants, nearly all of them said that they made initial contact through a personal friend who was already part of the movement.

For example, 'Ross' joined the Australian Conservation Foundation 'to gain access to conservation information'. On reflection, however, he added that he would not have joined without his close friendship with a person already active in the organisation. 'Michael' became an active environmentalist through his involvement with the Association of Australian Sustainable Communities through a personal tie with one of its members. 'Enid' became involved in collecting signatures for The Wilderness Society's Lindeman Island petition with a number of women living in her 'church units'. One of these women, 'Sarah', although not a formal member of The Wilderness Society, was very active in the society's affairs.

Just over 5 per cent of interviewees denied the existence of such a relationship, and insisted that it was their own personal motivation that was catalyst enough for their original involvement. This type of denial was also quite common in Gerlach and Hine's early and seminal study of the Pentecostal and Black Power movements and defies the norms of the American/European theoretical split mentioned above. In fact, the idea of movement growth through pre-existing social ties proved offensive to many. Gerlach and Hine (1970, p 87) reflected that:

> Participants, like many social scientists, prefer to develop ideological reasons for a movement's success. Social scientists focus on the type of appeal and its relationship to the particular type of deprivation or susceptibility they believe is causative. Participants suggest that the movement is spreading because more and more people are coming to accept the truth of their beliefs.

Once these links have been established and continued communication takes place, only then will a loose, collective and symbolic identity emerge, one that is not based on pre-determined, rational-comprehensive goals.

The environment movement in Australia is not based on a set of common goals. The diversity of the green symbol can only really be understood when one assesses the myriad different political theories that rally under its banner (see chapter 4). When we look beyond the myth of the common goal we are able to view the environment movement more accurately: as an amorphous and complex network of formal and informal social relationships. It is to the informal politics of specific networks and informal groups operating within the environment movement that we now turn.

NOTES

1 This initial interview schedule was carried out as part of research into the conservation movement in Queensland in 1985. The results of this survey were included in my article 'The Myth of the Common Goal: the Conservation Movement in Queensland', *Social Alternatives*, 1987, vol. 6, no. 4, pp 33–6. I have continued to survey environmental activists on an Australia-wide basis since the pilot research.

2 These two organisations do not sit at opposite poles. The heterogeneity of goals is even broader within the movement than the above example suggests, for these two organisations do have certain 'systems' goals in common. Their goals are both formalised to the extent that they belong to organisations that have constitutions. By consciously abstaining from involving themselves in formalised, hierarchical structures, other individuals, networks and informal groups active within this mass movement give it an additional dimension.

THE INFORMAL 'STRUCTURE' OF THE MOVEMENT

As network politics is so informal, the best way to access and under-stand its operation is to be involved or to be close to it, using a particular campaign or context, as there are no minutes with which to monitor the operation of networks, no formal records (though e-mail is changing this somewhat). With the networks that formed around the Howard government's moves to establish and implement the Environmental Protection and Biodiversity Conservation (EPBC) Act in 1999, for example, we have seen how networks can form and how important they can be. The informal network I characterised as 'the gang of four' made key decisions, allocated resources, and exercised power. Each member was an executive officer of a mainstream 'conser-vation' organisation and received the seal of approval from minister Hill and his Democrat offsiders. The Australian Conservation Foundation was excluded from the network because, as she quite can-didly admitted, Senator Meg Lees could not trust the organisation, as it had vociferously opposed her party's position on the GST.

The environmental issue chosen to illustrate the fluid and multi-directional communications structure of networks is the campaign to protect the wet tropical forests of north Queensland. This campaign included some of the most important environment actions carried out by key wilderness networks in this country. The longevity of this cam-paign (or series of campaigns) allows us to track the rise and fall of dif-ferent networks through time. Chapter 10 revisits this campaign, analysing its latter days and the emergence of a dominant network of professionals.

The wet tropics campaign was extremely diverse. It was directed at conserving all or part of the remaining rainforest in the coastal fringe of north Queensland. Since 1985 the term 'wet tropics' has been used to include such specific, geographically based subissues as Daintree, Downey Creek, the Windsor Tablelands and the Atherton Tablelands. The wet tropics issue—or collection of issues—had, however, far broader meaning. Its power lay in its ability to appeal to a vast range of people on many different levels. First, it served to focus the debate over the environment/preservation of natural forests and traditional forestry practices. Second, it raised the issue of state versus Commonwealth powers. Third, it was a wilderness issue, a question of the right of wilderness to exist independently of any rights humans might have. Finally, the building of the Cape Tribulation Road became a popular metaphor for a great issue confronting humanity: its continued existence.

Many different and separate networks came to the fore over nearly a decade, each one often advocating a different focus, a different direction, and involving different people. Table 2.1 (pp 26-27) outlines five separate attempts to define the network of groups involved in the campaign.[1] The first attempt, network 1, is the product of the Rainforest Conservation Society of Queensland (RCSQ), based in Brisbane. Network 2 was put together by the Wildlife Preservation Society of Queensland (WPSQ), also based in Brisbane. Both of these networks were defined in the first half of 1984.

Network 1 lists 15 groups that, according to RCSQ, were involved in one aspect or another of the rainforest campaign in Queensland. WPSQ lists 16 groups it regards as involved intimately with the rainforest issue. While the number of groups that make up these networks is similar, the organisational membership of the networks is quite different: the two networks have only nine organisations in common.

Networks in the environment movement differ markedly on most occasions, regardless of the fact that listings are often made with reference to the same subject at approximately the same time. The person who originally defines the network imposes his or her conceptual boundaries on the issue. In the case of both network 1 and network 2, only those groups operating in Queensland on wet tropics issues are listed. This list ignores inputs from interstate and overseas. It also lacks comprehensiveness for those groups that are most active at a local level. It reflects the biased perceptions of these two state- and city-based environment organisations.

Network 3 is composed of groups interested specifically in the Daintree campaign, as outlined by the Australian Conservation Foundation. The ACF is a national body, so it perceives the network as being Australia-wide. It does not, however, list organisations outside Australia that are also interested in the protection of

Queensland's and Australia's rainforest. Network 3 also lists those groups and organisations that are fundamentally interested in the Daintree road issue, as opposed to those with a broader focus on the wet tropics environment. To some environmentalists, however, Daintree is far more than just a subissue of the wet tropics campaign: it is the public focus of the entire campaign. According to Hayden Washington of the ACF and TWS, 'Daintree' is 'a key word in our national campaign to sell the wet tropics'. This network is defined by the perceptions of the organisation that creates it. The Wilderness Society and the Australian Conservation Foundation concentrated their resources on the Daintree road issue. They perceived this area as more ecologically important, and they also used this subissue to promote public awareness of the broader issue of the wet tropics. So this is not a wet tropics network, such as those produced by the two Brisbane-based organisations, but a Daintree network, which, though broader because of its national perspective, does not include those groups interested specifically in such issues as Downey Creek or the Windsor Tablelands.

THE ROLE OF NETWORKS IN THE WET TROPICS CAMPAIGN

Environment networks are primarily created to collate and disseminate information. Given the geographical dispersion of environment groups in Queensland and Australia generally, each of these groups needs to receive information that would not normally be available to them through channels of communication established by more formalised organisations, such as the Queensland Conservation Council (QCC) or the Australian Conservation Foundation. This distribution of information is essential. First, it helps to avoid duplication of effort. Second, when correctly digested, information is knowledge. With access to certain knowledge, environmentalists are less inclined to 'reinvent the wheel', and campaign strategies can be created with this collective knowledge. Finally, access to information provides groups within the movement with a sense of belonging to something far larger than themselves: a politically powerful social movement. This feeling inspires greater effort and achievement.

Information in the wet tropics campaign was disseminated in three significant ways: written correspondence, telecommunication, and personal contact. Through written correspondence, members of networks are kept up to date with personal and group opinion on such facets of the campaign as overall strategy and directions for the future. On occasions, letters are distributed to all members of a particular network, while at other times, as the material in question becomes more personal, they are sent to a more restricted group.

TABLE 2.1
FIVE NETWORKS IN THE WET TROPICS CAMPAIGN

NETWORK 1
As defined by the Rainforest Conservation Society of Queensland,
February 1984

1   Rainforest Conservation Society of Queensland (RCSQ)
2   RCSQ, Far North Branch (no official connection with RCSQ no. 1)
3   Cairns and Far North Environment Centre (CAFNEC)
4   Wildlife Preservation Society of Queensland (WPSQ)
5   WPSQ branches, especially Innisfail, Cairns
6   Townsville Regional Conservation Council (TRCC)
7   Queensland Conservation Council (QCC)
8   Wilderness Action Group (WAG)
9   Conondale Range Committee
10 Fraser Island Defenders Organisation (FIDO)
11 The Cooloola Committee
12 National Parks Association

NETWORK 2
As defined by The Wildlife Preservation Society of Queensland Inc.,
June 1984

1   The Wilderness Society, Brisbane (TWS)
2   Queensland Conservation Council
3   Rainforest Conservation Society of Queensland
4   Rainforest Conservation Society of Queensland (Townsville)
5   Tropical Rainforest Society (TRS)
6   North Queensland Conservation Council (NQCC; listed above as the
    now defunct TRCC)
7   Cairns and Far North Environment Centre
8   Australian Conservation Foundation (ACF)
9   Wilderness Action Group
10 Cape Tribulation Community Council
11 Wildlife Preservation Society of Queensland
12 WPSQ branches, especially Cairns, Innisfail, Tully, Hinchinbrook, Townsville

NETWORK 3
As defined by the Australian Conservation Foundation, 1984

1   Sydney
        a. The Wilderness Society
        b. The Colong Committee
        c. Australian Conservation Foundation

2   Melbourne
        a. Australian Conservation Foundation
        b. The Wilderness Society

3   Brisbane
        a. Rainforest Conservation Society of Queensland
        b. The Wilderness Society
        c. Queensland Conservation Council
        d. The Wildlife Preservation Society of Queensland

4   Hobart
     a. The Wilderness Society

5   Canberra
     a. The Wilderness Society
     b. Australian Conservation Foundation

6   Adelaide
     a. The Wilderness Society

7   Cairns
     a. Tropical Rainforest Society
     b. Australian Conservation Foundation
     c. Cairns branch, WPSQ

8   Queensland (others)
     a. Wilderness Action Group
     b. Australian Conservation Foundation
     c. Cape Tribulation Conservation Council
     d. Townsville branch RCSQ
     e. WPSQ branches at Innisfail, Tully, Hinchinbrook, Townsville

NETWORK 4
As defined by Dr J.G. Mosley (Director ACF), May 1984

1   L. Bourne
2   A. Keto
3   J. Sinclair
4   W. Huxley
5   R. McKendrick
6   R. Hill
7   G. McIntyre
8   D. Thompson
9   P. Figgis
10   A. Hingston
11   H. Washington
12   J. West
13   P. Ward
14   R. Scott
15   M. Wilcox
16   J. Staples

NETWORK 5
As defined by Dr A. Keto (President RCSQ), September 1984

1   Karen Alexander, TWS Melbourne
2   Jonathon West, TWS Canberra
3   Joan Staples, ACF Canberra
4   Aila Keto, RCSQ Brisbane

Source: Doyle 1986, p 28.

A more formal form of communication is the report or newsletter. One example of this in the wet tropics campaign was the *Wet Tropics Network—Report No 1*. This report attempted to provide a more regular and well-defined avenue of communication between groups involved in a particular wet tropics network. Each group stated in this brief résumé what it was doing to achieve the network's overall objectives. When possible, they mentioned plans for the immediate future, and invited comment from other network members. Unfortunately, this type of communication was not frequently practised in the wet tropics campaign.

Due to the unique political culture in Queensland, with an unusually scattered population, telecommunications took on added importance. Phone link-ups between network members occurred regularly in the campaign for discussion of resource allocation and strategy. An example is one organised by RCSQ in January 1984 including representatives from Cairns (Cairns and Far North Queensland Environment Centre), Townsville (North Queensland Conservation Council), Brisbane (Rainforest Conservation Society of Queensland), Sydney (TWS), Melbourne (ACF), Canberra (TWS), and Hobart (TWS). Given the geographical separation of eco-activists and a lack of resources, it is not very often that all members of a particular network can get together. Such meetings are often labelled 'strategy meetings'. Two such meetings took place, at an 'official' national level, in 1984. Both were primarily organised by the Australian Conservation Foundation to discuss such topics as the focus and the leadership of the campaign.

NETWORK COMMITTEES

Networks develop at a level involving all those groups interested in a particular issue. These networks are often too large to produce effective decisions. Member organisations involved in the network are often geographically separated from one another. It is necessary to divide these issue-oriented networks into smaller, committee-sized groups, which in turn create separate networks. These committees seem no longer to reflect organisations involved in the campaign, but are comprised of individual eco-activists. Again, these network committees differ in membership, depending on who initially defines them. That person's biases sometimes influence the eventual membership of the network. For example, certain people who appear to lack expertise and representativeness are included in committee-style networks because they are connected with the person defining the network.

Individuals may also excluded from an information network because they are out of favour with that key person. One possible example of this is found in correspondence between Hayden

Washington and Aila Keto, president of RCSQ (Doyle 1986, pp 27–33). It seems that Washington was excluded from the wet tropics campaign network, as Keto defined it, because the two did not have a good personal relationship. Such exclusions are explained as 'inadvertent slip-ups', but they happen far too often to be inadvertent.

It is often a conflict of interest that leads to a particular person being excluded from a network. For example, objectives and/or strategies may differ quite markedly between the definer of the network and the person seeking membership. Often there is no philosophical reason that justifies an omission and it may simply be a matter of a personality conflict. In some environment groups and networks, sexual relationships and fallings-out play a big role in determining who is included or excluded. Seldom is this openly acknowledged.

To express interest in a particular issue and enjoy the favourable opinion of the creator of the network in question, though important, does not guarantee membership. Networks are dominated by those who have attained status, expertise and experience. Status is commonly attained through successful performances on past campaigns, though this does not always apply. For example, a great deal of prestige is attributed to academic 'experts' and to those in the legal profession. Though less overt, traditional forms of dominance, such as male domination, are also evident. Once we observe these smaller networks committees, we begin to discard much of the pseudo-formalisation that occurs in larger collective forms.

Network 4 lists only names of individuals, and does include the organisations to which they are affiliated. It was compiled by Geoff Mosley, then director of the ACF, and was distributed to members of the network as he defined it. It is his, and to a lesser extent the ACF's, perception of the key individuals involved in the Daintree campaign. Of course these individuals were involved in some way or another with different organisations, but they were not representative. Often, two or more of these politically powerful individuals were affiliated with the same organisation, while other organisations were not represented. It is with this collection of individuals that the ACF discussed such critical facets of the wet tropics or Daintree campaign as new directions and coordination of strategy.

These committee-style networks range in size from about 12 to as few as four. Network 5 lists four prominent eco-activists in Australia. This committee was proposed in September 1984, partly due to the apparent failure of the larger, more representative networks to coordinate environment activities in relation to the wet tropics. Aila Keto drew up this 'small committee' as, in her own words, 'it is apparent that these is little co-ordination within the [wet tropics] campaign. Decisions appear to be taken on an ad hoc basis.

We need a carefully planned and clearly defined strategy' (quoted in Doyle 1986, p 29).

This committee never received recognition from environmentalists in Queensland or at a national level. At a special general meeting of the QCC in July 1985, a session was devoted to the wet tropics campaign. The proposed committee drew criticism, as participants believed it was not representative of the environment voice in the wet tropics campaign. Haydon Washington was critical of Keto:

> I certainly do not see the solution to these differences to lie in the creation of a 'strategy committee' that conveniently leaves out people of different views from yourself, no matter how actively involved they are in the Daintree campaign, as I am in ACF and TWS. The Daintree network exists, as many people around Australia are trying to get on with the job. If you think the network needs to discuss strategy more, by all means propose something. (quoted in Doyle 1986, pp 30–1)

The real issue here is the struggle to obtain political power, to define the framework of policy-making. No doubt Keto believed that to reduce the size of the wet tropics networks would also be directly paralleled with increased political effectiveness and efficiency. The problem was that those no longer participating would lose political power, while the select few would gain more. This is an excellent example of why the initial definition of networks taking part in a particular campaign is so important. It decides who will receive information and who will make the major decisions. Network 5 never received approbation from the environment movement, so the attempt to legitimise it failed. This does not indicate that the network ceased to exist, but rather that the role of these four powerful individuals and the status of this network remained informal and concealed. This network is remarkably similar in its size, nature and political process to the aforementioned network that was to form in relation to the EPBC act some 15 years later. Cam Walker (1999, p 2), national liaison officer of FoE, comments on the more recent committee:

> At no stage were WWF, HSI, TCT, or QCC given a mandate to represent the environment movement in negotiations with political parties over the EPBC Bill. At no stage have the representatives of these groups attempted to consult the rest of the network: in fact there is evidence that consultation processes were deliberately ignored.

Other networks emerged and then subsided throughout the wet tropics campaign. Another network of professional environmental activists (known here as Network 6) came to the fore in the wet tropics campaign in the late 1980s.

PERSONAL RELATIONSHIPS

Despite the fact that the politics of the wet tropics became increasingly institutionalised, informal politics always flows beneath and above the formal surface. Another level of informal relationships is the 'circle of friends'. It is difficult to discover the dynamics of such circles of friends without intruding on the participants' personal lives. Yet it is necessary to mention these relationships because it is at the personal level that crucial decisions are made that, later on, affect the formation of the agenda that is eventually acted upon in the more public forums.

The importance of personal relationships is accentuated in the environment movement. Because of the relative youth of the movement and the consequent failure to define and then formalise approaches to environmental issues, eco-activists rely heavily on an informal and personal trading of ideas. These relationships fill a void. A staff member of the Council for the Protection of Rural England, commenting on the multitude of personal contacts between environmentalists in England, expresses sentiments similar to those of his Australian counterparts: 'environmental groups work closely together. There is extensive co-operation. We work as a team. We all know each other, drink together and are personal friends' (cited by Lowe and Goyder 1983, p 82).

There are several reasons why the movement has not established more formal relationships. One relates to the largely autocratic frameworks of most formalised structures in Australia's environment organisations. Decisions are often made at the professional executive level, with little input from the membership. There is a lack of formal communication within and between organisations. This is why so much emphasis is placed on the 'circle of friends' as a political forum. This has been the most important and effective level of communication in the sphere of problem-specific policy creation. This high degree of informality assists the movement to maintain its flexibility, which is a necessary trait of the politically active in a field of endeavour that is constantly evolving.

PROBLEMS WITH NETWORKS

Some eco-activists hold basic misconceptions about networks, which ultimately affects their efficiency. The network enables the movement to be arranged in temporary categories. Categorisation is based on the prime interests of eco-activists. In this case these interests related to environment issues in the wet tropics. Thus the philosophical and political diversity of groups and individuals involved in a network is not as evident as it is in the movement as a whole. Often the focus of a particular individual or organisation reflects a specific political ideology. This allows the participants in the network to foster a productive working relationship.

The network temporarily provides the movement with a degree of consensus impossible to ascertain beforehand. The environment movement has practised consensus decision-making successfully on a number of occasions. For example, in the wilderness surrounding the Franklin River in south-west Tasmania (see chapter 8), consensus decision-making was exercised to such an extent that it became a philosophical tenet of the campaigners involved. There were several unique circumstances that allowed eco-activists in Franklin campaign networks to maintain a high degree of consensus. Decisions had to be made swiftly and economically. In the wet tropics campaign in Queensland, there were also occasions when decisions had to be made on the spur of the moment, such as those imposing the two blockades of the Daintree road in 1983 and 1984 (see chapter 3). No doubt trade-offs and compromises were made that in more normal times would not have been conceded. Consensus works most effectively within the environment movement at a time of crisis.

Still, enforced consensus should only be sought when necessary, for the philosophical diversity of the movement is one of its strongest attributes. Bruce Davis (1981) supports this view:

> Like the community from which it is drawn, the Australian environment movement consists of people of various ages, diverse political persuasions and different personalities. In this blend of experience and idealism lies the strength of the movement.

Unfortunately, the sense of unity provided by the existence of temporary networks surrounding specific issues (and strengthened by a sense of crisis) encourages environmentalists to assume that such a situation is the norm. In turn, resources are misspent on trying to develop a strategy that would be conveyed to all groups involved with a request that their actions be guided by that strategy and the underlying policy.

To establish such a steadfast, overall campaign strategy is an attempt to formalise these network relationships in a bid to capture their communicative essence, and thus allow these decision-making avenues to be repeated at will. This is not possible, as networks cannot be mapped onto each other. They are not informal groups, formal organisations or institutions. This approach does not acknowledge that the flexibility and unrepeatability of temporary networks can be a huge advantage (Piven and Cloward 1979). Scarce resources would be far better spent on distributing information, which would in turn promote unity and perhaps a flexible working strategy. Laying down hard and fast rules at the beginning of an environmental campaign is a destructive practice that only aggravates other participants, who have their own ideas about how to run a campaign.

Although some common lessons have been learnt, not all environment campaigns are based on the same philosophical foundations. Since networks are a collection of subissues, it is impossible to transfer a network established around one issue, with its particular set of contributors with divergent political ideals, to a new one. This was not understood in the wet tropics campaign. According to officials in The Wilderness Society, their involvement in the Daintree campaign—and their later participation in the broader theme of the wet tropics—was necessary if the temporary structure that was successful in the Franklin campaign was to be maintained. So the Tasmanian Wilderness Society became The Wilderness Society, and the primary interest of its active membership was redirected away from the Franklin River and its surrounding wilderness to those issues relating to the wet tropical forests of Queensland (see chapter 6). With the transfer of a previously established network to a new issue, the network no longer existed as a series of temporary channels of communication, but began to resemble a more formalised structure: an organisation.

INFORMAL GROUPS: FIGHTING THE PRESSURES TO FORMALISE

Informal groups maintain some elements of the informal style of politics inherent in networks. Most obviously, their internal relationships have not been defined by a constitution. They are, however, slightly more permanent than networks, displaying a group name, which provides the collective with a more tangible identity, however short-lived. These groups sit between the network and the organisation on the informal–formal continuum. The pressures to formalise are many, but some groups manage to resist these pressures in their pursuit of more radical and democratic goals.

For every formal organisation there are many more informal groups active in environmental politics. The first thing that is apparent about the majority of informal groups is the dominance of women. Although women dominate the non-institutional sectors of the environment movement, including NGOs, it is in this political milieu that women almost exclusively provide leadership roles. There are number of reasons for this. Two Canadian researchers have argued that women are more capable of eliciting localised support than men because of their central involvement in this local level of human political existence (Smith and Freedman 1972, p 170).

Whatever the explanation, at the informal level the movement is predominantly female. Many of these networks and groups challenge the politics of patriarchy and seek to promote a more equitable and far-reaching polity. I have described elsewhere the traits of these groups:

Although a group is more permanent than a network, it remains fluid and temporary in legal and established political terms. Persons operating within these groups openly advocate the continuation of what they construe as the open, non-hierarchical, non-institutionalised 'structure' of the group. Also, they believe that it is necessary to persist with the informal, voluntary form of human collective behaviour to achieve more radical, sometimes revolutionary, environmental change … These groups wish to remain structurally open, and a conscious effort is made on behalf of all [their] members to encourage and build a participatory democracy, with ground rules created only when deemed absolutely necessary. (Doyle and Kellow 1995, p 111)

Informal groups can be divided into two categories: introspective groups and non-introspective groups. The latter are often not conscious of their political form, their structure or their ideology. These groups often emerge as a result of some crisis in a particular locality. Often they are derogatorily referred to as NIMBY (not-in-my-backyard) groups. They exist everywhere. Adelaide has seen an explosion of non-introspective groups since the mid-1990s, though they have always been around. In October 1999, for example, a range of these 'action' groups gathered in a mass protest for 'environmental democracy' organised by the groups themselves, and facilitated by the Conservation Council of South Australia. The groups included the Campaign Against Pelican Point (CAPP), which formed to oppose the building of a power plant by British giant National Power on the Port Adelaide River; the Mount Barker Clean Air Group, opposed to the pollution emanating from the Mt Barker foundry; the Whittle Action Group (WAG), formed to fight the building of a skate park too close to residents' houses; the Kanmantoo Anti-Pollution Group; the Dumps Coalition, fighting the disposal of land-fill waste in residents' neighbourhoods; the People's Environmental Protection Authority (PEPA), formed in protest over the Environmental Protection Agency's appalling performance in South Australia, most obviously its failure to act promptly during the Mobil refinery oil spill at Port Stanvac; and the Eden Valley Native Vegetation Group, formed in opposition to wine giant Mildara Blass's application to clear 200 old native trees.

These action groups genuinely emerge from the community. They do not form in response to programs such as Bushcare, Coastcare, Landcare, Trees for Life. I prefer to refer to the latter as *bureaumunity* and *communeaucracy* groups rather than community groups, since they are a government-initiated hybrid community form. Many of the genuine community-based non-introspective groups disband when the immediate threat or issue has been decided. Some do go on to become formalised organisations, but this is not the norm.

In Kellow and Doyle (1995, chapters 4 and 8) I dealt with their

The campaign against Pelican Point in Port Adelaide galvanised many people who had not previously been involved in environmental politics. *Aussie Kanck*

A march for 'environmental democracy' beginning on the steps of South Australia's Parliament House, October 1999. *Aussie Kanck*

characteristics in some detail, and attempted to bring this form of political structure to life in a detailed case study based on the Hemmant Action Group (HAG). I want to focus here on introspective groups, which are perhaps less numerous than their non-introspective cousins. Introspective groups play politics outside the legitimate realm, and their dabbling in the underworld of illegality avoids the gaze of 'normal' analyses. This invisibility does not mean that they are not important. The activities of these groups often go unreported by the mainstream media, which prefer to pick up on the NIMBY actions, with plenty of colour, movement, and self-interest, which journalists can understand, as well as on the press releases of mainstream green organisations.

Most significantly, introspective groups challenge the dominant notion that ever-increasing formalisation is 'normal' and 'inevitable'. In their attempts to refrain from formalisation, and their belief in the importance and maintenance of a flexible, dispersed politics, they have had a momentous influence in the decentralised structure of the environment movement. It is this ideological element that has largely prevented the formation of a national umbrella organisation and for a long time thwarted attempts to form a centralised green political party.

The pressures to formalise human relationships can be intense. I want to explore briefly two sets of theories largely derived from the writings of Max Weber and Robert Michels. These theorists explain what they see as the inevitable transition of radical, democratic groupings into oligarchies and heavily bureaucratised, hierarchical organisations. Weber believed that human efficiency was compromised by the ability of any one individual to unfairly use his influence to achieve personal goals. He proceeded to outline an instrumentality that would effectively outlaw irrationality, emotion and incalculable humanism from administrative operations. He termed this process 'the demystification of experience' (Weber 1958). Weber argued that a system of societal administration based on bureaucratic rationality and 'demystification' would become the dominant human interactive form of the twentieth century. This form of political relationship is called 'bureaucracy'. Katharine Newman (1980, pp 143–64) describes the Weberian perspective as follows:

> In contrast to 'patrimonial' forms of administration, bureaucracy encouraged and indeed required that predictable, dependable methods of goal attainment be instituted. Rational rules replaced the whims of ad hoc decision makers; qualifications for positions developed where entirely ascriptive characteristics had previously prevailed; circumscribed, hierarchically distributed responsibilities took the place of kin-based or personalistic obligations; in short, rationalization as a 'master trend' in history was manifested in the increasing accountability which bureaucratic organizations reflected.

Weber further contended that, given time, informal networks and groupings operating within social movements would inevitably transform into more formal organisational types. This sociological journey was fundamental to his writings. This transformation would be coupled with a great change to the participants' original goals. These changes may include goal diffusion, which transforms definite, specific goals and ultimately allows the organisation to pursue goals from a broader range. These goals, Weber pointed out, become increasingly conservative, and ultimately the goal of organisational maintenance becomes all-consuming (Gerth and Mills 1946). Chaplin and Tsouderos (1956, pp 342–4) also write from a Weberian perspective:

> Accordingly, the process of formalization was interpreted to imply a sequential, stage by stage development of voluntary associations over time, an increasing complexity in the social structure, a progressive prescription and standardization of social relationships and finally, an increasing bureaucratization of the organization.

Similar ideas are found in the works of Robert Michels. Michels (1962) wrote *Political Parties* in 1911 when central Europe dominated political and sociological academic thought. Europe was about to embark upon two wars that would pose questions pertaining to the very nature of humanity. In hindsight, it may be argued that it was quite understandable for Michels to give a pessimistic review of democratic structures, particularly the focus of his empirical research—the German Social Democratic Party.

Michels argued that power in modern society inevitably concentrates in the hands of a small minority, no matter how democratic the intentions of the participants. His work questioned Marx's vision of a classless society and, as T.B. Bottomore (1977) argues, 'it formed part of a political doctrine which was opposed to, and critical of, modern democracy'. Marvin Olsen divides Michels' pessimistic view of the futility of democracy and the inevitability of elites into three categories: 'structural', 'operational' and 'practical'. Olsen (1970, pp 108–9) summarises the first category as follows:

> His structural arguments were based on the fact that virtually all organizations, from labor unions to total societies, tend to have a hierarchical structure in which authority and other forms of power are exercised downward from the top. Even though final authority may be vested in the membership as a whole, the necessity for leadership and over-all coordination makes popular voting and similar procedures mere rituals. Instead of being servants of the members, leaders become in fact the legitimate rulers.

Olsen's second category is based on the premise that leaders share many 'operational' advantages over ordinary members. For example, over time, elites acquire access and control over information, money, the agenda and special skills. This elite's control over the organisation increases, according to Michels, when voluntary workers become professional in the supposed interests of 'efficiency'. Increased efficiency is also an argument used to stamp out 'internal critique'. Michels (1962) concludes that eventually elite, professional control over the masses becomes so strong that the leader is eventually perceived as 'indispensable': 'One [who] is indispensable has in his power all the lords and masters of the earth.'

Michels argues that there may be no practical alternative to oligarchy. Leaders protect their own positions of power in an effort to see their own policies come to fruition. Their right to do this is rarely questioned, as most ordinary members do not have the time or interest to wield this power and to establish more representative policies (Olsen 1970, p 109).

Michels attracted much criticism due to the all-consuming nature of his 'Iron Law of Oligarchy'. He reflected the European tradition of the time of attempting to create grand models of society: laws that would account for all human behaviour. Several critics have merely dredged up separate examples that cannot be squeezed into the confines of Michels' model. While proving that not all experiences can fit into any one model, these scenarios do not prove that Michels' elite theory cannot be applied at all.

One of the most articulate challenges to Michels' position comes from John Plamenatz. In *Democracy & Illusion*, Plamenatz argues that Michels' work fails to distinguish between political organisations and political systems. To Michels, larger human collective forms such as social movements operated under the same mechanics as formal organisations with a constitution. Plamenatz (1973, p 105) questions this assumption when he writes:

> All that needs to be established by argument is, he thinks, that the iron law does operate, even in popular and proletariat organisations. But this, so it seems to me, is a mistake, for a political system, though it includes organisations, is not in itself one, at least not in the same sense as what it includes. We cannot simply take it for granted that it must be undemocratic if important organisations indispensable to its functioning are so.

I would argue that in many democratic structures, once organisation becomes apparent, forces may come to the fore that invite oligarchy. This trend is far more flexible than Michels' insistence on an 'iron law'. This difference in degrees of acceptance comes about for a variety of reasons. Michels was engrossed in rigid 'grand models' of society, whereas I operate in a far more eclectic intellectual era. Michels was also an unshakeable empiricist.

Oligarchy did occur within two formal organisations associated with the wet tropics campaign—the ACF and TWS—but not in most of the other groups that participated in that and other environmental actions. This supports Plamenatz' criticism of Michels that an organisation is a microcosm or part of a political system—or in this case a mass social movement—not an exact miniature of the whole.

So before we investigate the emergence of elite power, we shall look briefly at some introspective groups: groups that have *not* become oligarchic, and consciously maintain their informality in a bid to pursue more democratic and radical goals. Earth First! originated in the USA but is also at times active in Australia. It describes itself as a 'non-organisation' for these purposes. It has no constitution or any other legal or political forms that make it 'legitimate'. Indeed, one of the key reasons it remains illegitimate is that it perceives the 'representative democracy' of mainstream society as 'a sham, controlled as it is by the true criminals—corporate devils and government co-conspirators—who rape the land with impunity' (Taylor 1991, p 262). Earth First!, like most informal introspective groups, pursues strategies for environmental change outside the state, arguing that the bureaucratic and corporate mechanisms of the state are structurally responsible for most environmental degradation. As a consequence, lasting environmental solutions cannot be found within these structures. So rather than lobbying governments or working with the corporate sector, Earth First! involves itself in direct mass mobilisation and education programs, and in some very visible direct action campaigns. Some of the latter are the focus of the next chapter.

The Nomadic Action Group (NAG), an Australian group, has no formal organisation or planning, no set decision-making structure, no set aims or tactics, no agreed spokespeople or media position. NAG's goals are long-term and radical: it demands a complete halt on logging of hardwood forests in Australia (NAG is also dealt with more fully in the next chapter).

One group, the EcoAnarchoAbsurdistAdelaideCell?! (EAAAC?!), is an 'ideal type' of introspective group, in that it is an extreme example. Sustaining its diverse and anarchistic political form seems to take primacy over most other pursuits.[2] It is difficult to really know the parameters of its ideology, for, like NAG, it has no agreed spokesperson, no unified strategies, and no regular meetings. In addition to media infiltration, its actions are diverse, numerous and quite often secretive. In a letter addressed to the newsletter *Entelechy*, informing the editor of a takeover bid, an anonymous affiliate of the group describes of some of its strategies and pursuits:

It is impossible to ascertain exactly how big EAAAC?! is. Their actions are diverse, and the wider membership is constantly changing. They involve themselves in both violent and non-violent, direct and indirect actions … They believe (often early in the night of their spontaneous meetings) that the only way to true environmental change (thus averting the environmental crisis that several of them don't believe exists anyway) is through uncoordinated, diverse, and absurd political actions. (Anon 1993, p 2)

Table 2.2 lists 30 loose objectives with a social, ecological and absurdist flavour.

TABLE 2.2
ECOANARCHOABSURDISTADELAIDECELL?! (EAAAC?!): LIST OF OBJECTIVES AS DEFINED AT THE EXETER HOTEL, ADELAIDE, 11 SEPTEMBER 1996

1   Televisions and poker machines out of all public houses now.
2   Immediate moratorium on all native hardwood logging in Australia.
3   Carpet bowls to be introduced as an exhibition sport at the Sydney Olympic Games in the year 2000.
4   Halt of all demolition of heritage buildings until Guy Fawkes Day.
5   Farm the city squares.
6   Every second tree planted in urban areas must be a fruit tree.
7   Subsidies for business folk to purchase and wear clown suits.
8   North Terrace to be terraced.
9   Government and the judiciary—no lawyers need apply—to be selected by demarchy (by lot).
10   24 hour bus, train and ferry services.
11   Bicycles to be provided for people on every third street corner (with baskets).
12   No kikuyu or cooch grass on 'nature-strips'.
13   Community appointed editors (on a rotational basis) of the newsletter the *Advertiser*.
14   The construction of a 'crying wall' and 'laughing stocks'.
15   The construction of a museum of 'fuzzy logic' and 'economic irrationality'.
16   Set the animals free.
17   87% of arid land habitats to be declared 'parks' or 'non-representative reserves'.
18   Ban the corporate form of human interaction.
19   Ban uranium mining and exports.
20   Subsidise job share.
21   30 hour week maximum in the workplace, and no overtime.
22   The establishment of a 'flexitime secular sabbath'.
23   Hemp legalised.
24   Extinguish pastoral leases.
25   Non-indigenous people to apologise and to 'ask permission'.
26   Organic food only: no imports.
27   No guns.
28   Always remember the children.
29   Delete the monarchy.
30   Siesta reintroduced (2 p.m. to 6 p.m. in summertime).

An action by the EcoAnarchoAbsurdistAdelaideCell?!!. *Phil Bradley*

These objectives are not the equivalent of a constitution, because they are simply a list of non-agreed upon principles discussed at a single spontaneous meeting. They are useful, however, in pinpointing some of the key interests of some participants. The emphasis on the absurd seems to be one principle of EAAAC?! Even the political form of the group is taken to an absurd level of abstraction. Indeed, even the group's acronym has some comic dimension. It is actually pronounced '"eek" as in mice plague' (Anon 1993, p 2). The writings of two anonymous members of the 'core cell' focus on this absurdist dimension. The following is taken from the correspondence relating to the approach to *Entelechy*:

> They sometimes argue that it is only shocking and sick humour which is capable of knocking the overly serious, pessimistic and eternally repeating record of mainstream human history off its axis. In effect, it uses the absurd to shake the foundations of the accepted and environmentally nice ways of behaviour … For example, they demand a complete halt to all building in Adelaide and the total deconstruction of currently existing constructions. Many demolition jobs of housing in Adelaide by EAAAC!? have gone unnoticed because these actions have been disguised as home renovations. In fact, some EAAAC?! cell members often dress up in the garb of the home handy person. (Anon. 1993, pp 2–3)

This next piece comes from another letter warning of a separate takeover bid for the small newsletter *Megazine* three years later:

> The constipated are at the root of much that is evil in the world today. Bureaucracy operates on a symbiotic relationship with constipation, economic rationalism is a euphemism for constipation ... I speak as one who knows: yes, I confess it, I was once a member of the constipated cabal myself. I know the intense frustration, discomfort and irritation which governs their lives, and makes them hate the world and wish to obstruct the development of the good, the right and the true. (Anon 1996, p 1)

With 'absurd' environmentalism, spontaneity and open participatory democracy seem to be important to the group, though it is sometimes difficult to fathom what is a political action of EAAAC?! due to the group's dispersed and fluid nature. In a brief ethnographic study of this subculture, Adam Simpson (1994, p 1) suggests that actions are only identified with the group after they have occurred. He describes what he perceives to be the criteria the group uses to judge whether a particular eco-action can be identified with the EAAAC?!: 'spontaneity beyond what would be expected of any reasonable human being, activity should have made a considerable contribution towards the principles of environmentalism as a whole, activity should be non-reproducible, and activity must be performed while maintaining both a sense of humour and the absurd'.

EAAAC?! also maintains its amorphous form to confuse opposition, and to remain invisible to its political enemies. One commentator describes this political form as 'deliberately post-modern' (Darley 1996). There has been much political infighting recently, with one section of the 'core cell' arguing that the other four sections are 'not funny enough'. There are also tensions between those in the group who advocate indirect non-violent actions and those pushing for direct violence (Baker 1997).

EAAAC! is particularly absurd but its amorphous structure was adopted by another introspective group, the Kumarangk Coalition, this time for far more serious reasons. Since 1996, a plethora of legal actions have been aimed at individual environmentalists, groups and organisations opposed to the building of the now infamous Hindmarsh Island bridge in South Australia. It is normally argued in legal circles that to be incorporated as a formal organisation provides members with some degree of legal protection. The Kumarangk Coalition has opted for another structure. In refraining from legal incorporation and from creating a constitution, the group (for it is not a 'formal' organisation) has no 'members' as such, and remains a moving, sometimes invisible, target. Although many individuals and formal organisations are caught up in lengthy and costly legal proceedings against the proponents of the bridge—the Chapmans—the Kumarangk Coalition, with a diffuse leadership and decision-making structure, cannot even be served with a writ. The case of the Kumarangk Coalition now appears in findings of the District Court and Supreme Court of South Australia. The legal status of non-incorporated associations, and their 'non-membership', have set some fascinating legal precedents relating to informal groups.[3]

In many ways groups such EAAAC?!, NAG, Earth First! and the Kumarangk Coalition are the underground part of environment-based social movements, and it is extremely difficult to gain access to information. They remain a vibrant, though sometimes off-beat, sector. It is also difficult to 'measure' their worth as they do not always operate with specific goals in mind or according to administratively rational principles. There is no doubt, however, that they do serve a purpose, if it is only to make other environmental networks and NGOs appear more 'reasonable' to the wider community. It is refreshing to see clear objectives listed without a concern for political expediency. Too often political bargaining within formal organisations and institutions occurs when objectives are being developed, not afterwards. The result is often watered-down, incremental goals, with the lowest common denominator having already been moulded to fit within existing frameworks.

Another advantage of this style of politics is that many people who are no longer willing to place their energies into more traditional political pathways find inspiration, hope—and even humour—in these more

This poster symbolises the long battle of the Ngarrindgeri people against the construction of the Hindmarsh Island – Kumarangk bridge. *Aussie Kanck*

anarchistic 'connectives'. The informal parts of the environment move-ment in Australia have not been sufficiently studied or appreciated. The politics of networks and informal groups is, in many ways, what makes a social movement. Diani (1992, p 18) writes:

> Only the study of the properties of inter-organisational and interperson-al networks is, in this perspective, directly relevant to the analysis of social movements. In contrast, for example, the study of individuals' commit-ment to a specific movement organisation, albeit of obvious substantial interest, is not specific of social movement studies.

Just because some human relationships become institutionalised does not mean that the power of informal politics subsides. Informal politics remains an undercurrent in all political relationships, including those within and between formal organisations. Unfortunately, when informal politics is swept aside by a constitution, it can re-emerge with unpleasant characteristics.

We now turn to one of the most conspicuous forms of informal political activity: direct action. This is often pursued by networks, groups and organisations when, among other justifications, the 'legiti-mate' pathways defined by the state no longer appear adequate for some environmental goals. We look also at tensions within the move-ment between non-violent and more militant direct action.

NOTES

1   I first described these networks in Doyle 1986. I uncovered these networks through my involvement in the campaign, which allowed me access to correspon-dence that detailed the members of each network. For example, on certain occa-sions, the person who had defined a given network specified who should receive a copy of correspondence. Common information for network members is often pro-vided first to those who would ultimately be given the power to make decisions.

2   Like many introspective groups, EAAAC?! tends to associate itself with the anar-chist principles of 'social ecology'. For overviews of this environmental political philosophy, see Eckersley 1992; Bookchin 1987; and Doyle and McEachern 1998.

3   For legal reasons, and for the sake of clarification, I need to say here that the Kumarangk strategy of maintaining its informal structure may or may not have been intended, as I was not privy to these strategic decisions. In true McCarthyist style I have to say: 'I was never a member of the Kumarangk Coalition, nor do I know anyone who was a member.'

DIRECT ACTION IN ENVIRONMENTAL CONFLICT: A RE-EXAMINATION OF NON-VIOLENT ACTION

Any individual or network can embark upon direct action. It needs no formal approval from any organisation, corporation or institution. It often occurs as a last ditch attempt to protect a specific environment. It sometimes involves breaking the law. It is often the most obvious and visible form of informal political activity. In their book on one form of direct action, 'monkey wrenching', Foreman and Haywood (1987, pp 14–17) explain the advantages of this style of politics in wilderness conflicts in the USA:

> The effectiveness of conventional political lobbying by conservation groups to protect endangered wild lands will evaporate … It is expensive to maintain the necessary infrastructure of roads for the exploitation of wild lands. The cost of repairs, the hassle, the delay, the down-time may just be too much for the bureaucrats and exploiters to accept if there is a widely-dispersed, unorganized, strategic movement of resistance across the land … If loggers know that a timber sale is spiked, they won't bid on the timber. If a Forest Supervisor knows that a road will be continually destroyed, he won't try to build it. If seismographers know that they will be constantly harassed in an area, they'll go elsewhere. If ORVers [drivers of off-road recreational vehicles] know that they'll get flat tires miles from nowhere, they won't drive in such areas. John Muir said that if it ever came to a war between the races, he would side with the bears. That day has arrived.

Many of the beneficial characteristics that Foreman and Haywood attribute to their own form of direct action (see Table 3.1) are similar to certain traits of informal politics discussed in the previous chapter: deliberate non-organisation and non-formalisation, deliberate use of dispersed networks, and the articulation of the need for fun.

TABLE 3.1
SOME CHARACTERISTICS OF ENVIRONMENTAL DIRECT ACTION

'

• MONKEYWRENCHING IS NON-VIOLENT

Monkey wrenching is non-violent resistance to the destruction of natural diversity and wilderness. It is not directed toward harming human beings or other forms of life. It is aimed at inanimate machines and tools. Care is always taken to minimize any possible threat to other people (and to the monkeywrenchers themselves).

• MONKEYWRENCHING IS NOT ORGANIZED

There can be no central direction or organization to monkeywrenching. Any type of network would invite infiltration, agents provocateurs and repression. It is truly individual action. Because of this, communication among monkey-wrenchers is difficult and dangerous. Anonymous discussion through this book and its future editions, and through the Dear Ned Ludd section of the *Earth First! Journal*, seems to be the safest avenue of communication to refine techniques, security procedures and strategy.

• MONKEYWRENCHING IS INDIVIDUAL

Monkeywrenching is done by individuals or very small groups of people who have known each other for years. There is trust and a good working relationship in such groups. The more people involved, the greater are the dangers of infiltra-tion or a loose mouth. Earth defenders avoid working with people they haven't known for a long time, those who can't keep their mouths closed, and those with grandiose or violent ideas (they may be police agents or dangerous crackpots).

• MONKEYWRENCHING IS TARGETED

Ecodefenders pick their targets. Mindless, erratic vandalism is counterpro-ductive. Monkeywrenchers know that they do not stop a specific logging sale by destroying any piece of logging equipment which they come across. They make sure it belongs to the proper culprit. They ask themselves what is the most vulnerable point of a wilderness-destroying project and strike there. Senseless vandalism leads to loss of popular sympathy.

• MONKEYWRENCHING IS TIMELY

There is a proper time and place for monkeywrenching. There are also times when monkeywrenching may be counterproductive. Monkeywrenchers gen-erally should not act when there is a non-violent civil disobedience action (a blockade, etc.) taking place against the opposed project. Monkeywrenching may cloud the issue of direct action and the blockaders could be blamed for the ecotage and be put in danger from the work crew or police. Blockades and monkeywrenching usually do not mix. Monkeywrenching may also not be appropriate when delicate political negotiations are taking place for the protection of a certain area. There are, of course, exceptions to this rule. The Earth warrior always thinks: Will monkeywrenching help or hinder the protec-tion of this place?

• MONKEYWRENCHING IS DISPERSED

Monkeywrenching is a wide-spread movement across the United States. Government agencies and wilderness despoilers from Maine to Hawaii know that their destruction of natural diversity may be met with resistance. Nation-wide monkeywrenching is what will hasten overall industrial retreat from wild areas.

• MONKEYWRENCHING IS DIVERSE

All kinds of people in all kinds of situations can be monkeywrenchers. Some pick a large area of wild country, declare it wilderness in their own minds, and resist any intrusion against it. Others specialize against logging or ORVers in a variety of areas. Certain monkeywrenchers may target a specific project, such as a giant powerline, construction of a road, or an oil opera-tion. Some operate in their backyards, others lie low at home and plan their ecotage a thousand miles away. Some are loners, others operate in small groups.

• MONKEYWRENCHING IS FUN

Although it is a serious and potentially dangerous activity, monkeywrenching is also fun. There is a rush of excitement, a sense of accomplishment, and unparalleled camaraderie from creeping about in the night resisting those 'alien forces from Houston, Tokyo, Washington, DC, and the Pentagon.' As Ed Abbey says, 'Enjoy, shipmates, enjoy.'

• MONKEYWRENCHING IS NOT REVOLUTIONARY

It does not aim to overthrow any social, political or economic system. It is merely non-violent self-defence of the wild. It is aimed at keeping industrial 'civilization' out of natural areas and causing its retreat from areas that should be wild. It is not major industrial sabotage. Explosives, firearms and other dangerous tools are usually avoided. They invite greater scrutiny from law enforcement agencies, repression and loss of public support. (The Direct Action group in Canada is a good example of what monkeywrenching is not.) Even Republicans monkeywrench.

• MONKEYWRENCHING IS SIMPLE

The simplest possible tool is used. The safest tactic is employed. Except when necessary, elaborate commando operations are avoided. The most effective means for stopping the destruction of the wild are generally the simplest: spiking trees and spiking roads. There are obviously times when more detailed and complicated operations are called for. But the monkey-wrencher thinks: What is the simplest way to do this?

• MONKEYWRENCHING IS DELIBERATE AND ETHICAL

Monkeywrenching is not something to do cavalierly. Monkeywrenchers are very conscious of the gravity of what they do. They are deliberate about tak-ing such a serious step. They are thoughtful. Monkeywrenchers—although non-violent—are warriors. They are exposing themselves to possible arrest or injury. It is not a casual or flippant affair. They keep a pure heart and mind about it. They remember that they are engaged in the most moral of all actions: protecting life, defending the Earth.

Source: Foreman and Haywood 1987, pp 14–17.

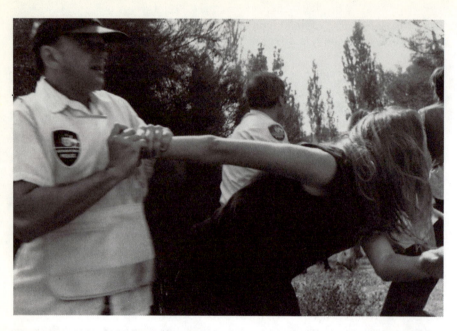

The long arm of the law. *Phil Bradley*

MDA

With increasing frustration at pursuing more 'legitimate' avenues for environmental change, there has been a worldwide increase in more militant, sometimes violent direct actions. In Australia, militant direct actions (MDAs) have been part the environment movement since major forest conflicts began in the late 1970s, but have definitely constituted a minority tradition. In recent times there has been a slight increase in the incidence of militant direct actions, but probably more striking has been the way non-violent action (NVA) has been reinterpreted by Australian environmentalists to include more forms of active resistance. On the whole, this has lessened environmentalists' frustration and increased the effectiveness of direct, non-violent eco-actions (see Doyle 1994a). I use the term 'MDA' in referring to any form of direct action that does not fit within the traditional rubric of non-violent action and specifically to actions that involve some risk to human beings.

The pressures to be more militant have always been there, often emerging as strategies in the evenings around campfire direct-democracy meetings where participants sought a consensus position for the next day's actions. For example, the Hungerford Sawmill was burned down at the peak of the campaign to save Terania Creek in northern New South Wales in 1979–80. This action was not condoned by the main body of protesters. Also at Terania, fallen logs were 'spiked' by an anonymous, breakaway group under the cover of darkness. Many local

Examples of MDA

environmentalists believed this to be a turning point in their favour, as loggers immediately stopped felling trees once the timber was worthless and dangerous to process any further.[1] In public, their position was still non-violent and these actions were not condoned.

Despite these sporadic, more violent activities, it must be made clear that non-violent action has dominated environmental direct actions during major campaigns. In Australia, non-violence was initially adopted by various groups and networks that advocated more radical change. A strong faction developed within the more radical domains of the early Wilderness Society. The theory behind non-violent action is that the ruler's power erodes when the people refuse to cooperate. This non-cooperation has traditionally taken the form of non-violent actions, including marches, rallies, pickets, strikes, boycotts, sit-ins and blockades.

One defining characteristic of Australian environmental direct action, whether non-violent or more militant, is that it has revolved around forest issues. Forest issues in Australia have largely defined the wilderness-oriented networks of the environment movement. The passion for rainforests is totally understandable. They are the repository of so many of the Earth's species and are awe-inspiring in their beauty. They are ideal places for direct actions. Protesters can conceal their whereabouts, and when conflict does occur, it is usually in a fairly confined space suited to media coverage.

THE BATTLE FOR THE DAINTREE

Traditional non-violent action dominated the goals and strategies in the battle for a section of rainforest in Queensland's wet tropics known as the Daintree, in what became one of the most famous forest conflicts in Australia's political history.

In November 1983 the Douglas Shire Council and the Queensland government attempted to construct a road from Cape Tribulation to Bloomfield. This road would dissect one of the last remaining and grandest stands of Australian rainforest. The proposed road catapulted the wet tropics onto the national agenda. The campaign, beginning with a blockade, unfolded quickly with little advanced planning. The participants included locals from the Cape Tribulation area (the Cape Tribulation Community Council was among these); committed people from other north Queensland centres, such as the Wilderness Action Group based at Port Douglas; and a group of individuals known as the Nomadic Action Group (NAG). Formal organisational input came from the Australian Conservation Foundation, whose wet tropics project officer was present, and the Cairns and Far North Queensland Environment Centre (CAFNEC). Participants clearly expressed typical motives, and their comments illustrated several forms of radicalism:

Greens challenge legitimate uses of force and coercion through non-violence. *Phil Bradley*

Beauty

Cape Tribulation is a really special place to a lot of people in Queensland. It is a wild place, an incredibly beautiful place. (Hill 1984)

Spiritual

I, too, have learnt how special a place it is. Since arriving here nine years ago, I have come to realise that it was in a forest like this that we began. This forest has cared for me like a mother cares for her child, she has fed me and protected me, calmed and healed my physical and mental wounds, and has taught me many things, including tolerance and forgiveness. (Hans, quoted in Wilderness Action Group 1984, p 24)

Wilderness

Where is the brotherhood we once shared with the trees? Where is the sanctity that we used to feel in the forests? Where is the respect for anything greater than ourselves? Forgotten in the plethora of images on millions of television screens. Expelled down the exhaust pipes of the swarms of motor cars that are apparently bent on colonising this planet. Submerged by the noise, frenetic pace and confusion of the concrete jungles through which we walk as strangers. (Grainger 1980)

Survival

The rainforests contain fully half of the ten million species of plants and animals on earth—the very womb of life. The continued evolution of life on this planet depends on the survival of the genetic materials that these forests contain. (Cohen and Seed 1984)

Aboriginal land rights—and sacred site desecration

The road will violate Aboriginal sacred sites. The Kuku Yalanji tribe have a number of sacred and cultural sites in the area. (CAFNEC, 1984, p 2)[2]

The spontaneous nature of this first Daintree blockade explains why these resonses were not finely honed strategic weapons designed to win over the collective minds of the Australian public. Rather, they were expressions of how participants felt about the threatened forest.

STRATEGIES

Some circle through the bush to get in front of the bulldozer. It ploughs straight ahead, into three people. Two dive out of the way at the last moment, but the third is caught and grabs a steel bar that pokes out of the front of the bulldozer. He is carried high into the air, clinging to the bar, and for long seconds his figure swings wildly against the dark green background of the silent trunks and leaves. A scream goes out as the others watch helplessly. Eventually that blockader returns safely to the ground. (Green 1984, p 26)

A blockade was in place with five or more people buried up to their necks, and barricades of posts, logs, banners and signs. People chained themselves to some of the logs ... Then John said he had a length of high tensile chain and wanted to be assisted up a tree, situated right in the path of the 'dozers, when they started work in the morning. He was dubbed high John, for the incredible heights to which he climbed in the tree. (Wilderness Action Group 1984, pp 17 and 47)

The strategies employed were militant and sometimes novel. There was no time to lobby politicians. Moreover, blockade participants were not convinced that mainstream party politics could protect the forests. No time existed, either, for carefully managed media stunts. Yet blockaders were scrupulously non-violent. A 'Blockade Information Sheet', hastily prepared by CAFNEC, spelt out to the second wave of prospective blockaders that the action must be 'totally peaceful':

Our argument is with the Douglas Shire Council and the Queensland Government, not with their agents on site, the workers and the police. We urge maintenance of peaceful and respectful relationships with these people. (CAFNEC, 1984, p 2)

The Wilderness Action Group wrote of the first blockade:

> The blockaders' reaction to this madness has been non-violent. This assurance has been given to police, Council and National Parks Officers … Rupert Russell, a well-known writer and naturalist, says that this action (bulldozing the road through the forest) is not in accordance with his principles, and is promptly arrested. He manages four steps before promptly collapsing limply to the ground. With his face in perfect repose, as though asleep, he is carried off to the paddywagon. (quoted in CAFNEC 1984, p 10)

A consensus process was used to select strategies:

> As the light fades from the sky, a large tarp is spread on the damp ground, the circle for the meeting begins to form, the chairs being quick to be occupied, till all is quiet and the agenda is passed around.
>
> The group is called to order by a chairman. Ideas start to flow, and pros and cons are aired, till someone says 'I'll do that!', then we are off again on another point … The meeting could have been over in half an hour but for one word … CONSENSUS. We ran every meeting by it, an extremely tedious operation … (Wilderness Action Group 1984, p 33)

Although time-consuming, this process is part of the radicalism of the resistance. Consensus represents the anti-hierarchical attitudes common in Australian direct action networks. For example, CAFNEC's information sheet states the desired distribution of power at these meetings: 'Meetings are held each night at 6.30 p.m. All people have an equal say. The blockade may be broken into smaller groups with elected representatives if it becomes too large' (CAFNEC, 1984, p 2).

If an individual held office in a particular organisation, she received no more kudos or power than any other blockader.[3] Even mainstream environmental leaders were accorded no deference. For example, the ACF's wet tropics officer begged protesters to halt the blockade as an agreement with the federal Minister for Arts, Heritage and Environment to cease logging was imminent. The protesters disregarded this plea. The first Daintree blockade was funded inexpensively. Despite the protesters' limited resources, their tactics forced observers to ask: 'Who were these people risking their lives for the forest? And what motivated them?'

THE NOMADIC ACTION GROUP

The Nomadic Action Group (NAG, offers an example of radical direct action in Australia during the 1980s. The media represented NAG as typical of the environment effort at Cape Tribulation: buried up to their necks in mud, or hanging precariously from trees (Green 1984, p 26). NAG 'has had a long history of direct action in eco-activity in

The Franklin River, now a symbol of modern Australian environmentalism. In 1983, this issue, above all others, was considered decisive in leading the ALP to power under Bob Hawke. *CCSA*

Australia' (Doyle and Kellow 1995, p 108). It has generated immense publicity for environmental issues and some of its direct actions have been ultimately successful (Terania Creek and the Franklin Dam[4] are just two blockades that achieved the desired result of all environmentalists: in the former a logging ban, in the latter the abandonment of a dam project). A brief history of the group reads:

> The group on Errinundra Plateau which varies in numbers from 25 to 80, was based around up to a dozen 'veterans' of the Nightcap, Franklin, Roxby and Daintree actions. Initially calling themselves the 'Nomadic Action Group' they encamped on the Errinundra around Christmas, having come from the Down to Earth Confest at Wangaratta ... The position of the most prominent in the group was that there should be a moratorium on all logging in Australia, an immediate halt on the whole Errinundra Plateau, and that only immediate blockade action was appropriate. Others did not necessarily agree. (Chris Harris, quoted in Doyle and Kellow 1995, p 108)

The operation of the Nomadic Action Group is dominated by one overriding tactic: non-violent action. Consensus is a principal focus, but other strands of philosophical and religious thought are also present. Since the Daintree campaigns of the 1980s have given way to other groups born of the NAG tradition, there has been a tendency to refer to this philosophy as 'deep ecology'. While deep ecology is now directly equated with bio- or ecocentrism, and constitutes a potpourri of ideas and values, in the late 1970s and early 1980s it was not a common term among direct action participants. Apart from the dominance of non-violent action, there was a wide variety of ideas. These more revolutionary ideologies depict the existing system as ecologically bankrupt. The revolutionaries, or radicals, advocate widespread political and/or spiritual change, while more mainstream reformists believe pragmatic changes within the system are sufficient to avert environmental problems.

The lessons of the first blockade at Daintree still dominate environmental direct actions in Australia. The experience further reinforced the validity of predominantly NVA tactics at both Terania Creek (New South Wales) and the Franklin Dam (Tasmania) blockades. All three campaigns were successful in the short term, with significant sections of forest being listed as World Heritage reserves by the United Nations, thereby strengthening the claims to adequate protection.

TENSIONS IN THE INTERPRETATION OF NVA: DIRECT ACTION IN THE 1990s

In the early 1990s it became obvious these areas were simply prized tokens, as the majority of Australian hardwood forests remained open to the timber companies. Poor long-term results led to frustration and increased incidences of direct action within the movement.

Media, police, environmentalists, action: this could be a snapshot of green protest anywhere in Australia. *Phil Bradley*

Some of these incidents were more militant, and included 'ecotage': 'secret acts aimed at stopping or slowing down processes judged as harmful to the environment' (Chandler, 1992). Many of these actions were derived from the list in Foreman and Haywood's *Ecodefense: A Field Guide to Monkeywrenching* (1987). These included: tree and road spiking, disabling forestry vehicles, slashing tyres, traplines, and removing and changing the position of survey stakes. In July 1990 contract loggers in south-east New South Wales estimated that they had 'sustained $630,000 worth of damage to their machinery, devastating families financially' (Chandler, 1992).

Another reason for an increase in the number of MDAs reported in the media may be that they were now carried out as 'media actions' having the express purpose of promoting public awareness of a particular issue are usually choreographed and less indicative of genuine last-ditch defence efforts. There were a number of these actions between 1991 and 1993. In Deloraine, Tasmania, for example, a train carrying woodchips was hijacked by environmentalists. These included a loose network of people, some Earth First! and TWS affiliates from the mainland, and local activists. As the rail tracks and the land immediately to either side were federal property, state police could not act. The protesters negotiated with police that if they helped organise media coverage of the event then they would leave peacefully. The Tasmanian police agreed to this request. The media arrived, the play-acting began,

These two photographs illustrate direct, minority actions. This event took place as part of the Westpac National Day of Action, the 'Westpac die-in', 28 September 1998. Westpac was targeted as it was the key financier to North Ltd, a major company involved in the construction of the Jabiluka mine in Kakadu, NT. *Jabiluka Action Group*

and when it was over, the protesters left. Over the next week, however, several of the protesters, having moved into some of the surrounding towns, were arrested on charges of vagrancy.

In Sydney in November 1992 the New People's Commission for the Forests (loosely affiliated with a number of groups including the Renegade Activist Action Force, the North East Forest Alliance, and Earth First! members attached to Friends of the Earth) was involved in an operation that 'took over' the Forest Commission's head office in Sydney. After seizing control of the building, the protesters used the Forest Commission's communications system to involve media. Representatives of the media were admitted to the building, where the anti-logging requests of protesters were broadcast. Two protesters were later charged and successfully prosecuted.

Similar though less militant media actions took place at protests on the Errinundra Plateau in south-east Gippsland in December 1993. Media coverage was arranged beforehand, the police were notified and so were forestry officials. A number of new protesters, who had arrived earlier in the week, were gathered together and taken to meet the cameras. The media collected their footage and photographs and then the protesters went home.

Such media actions are of limited value. There can be no doubt that they gain media attention, but hijacking trains and buildings seems unwarranted to most people. While they are of limited use as media-catching devices, there are more direct negative consequences, such as the total loss of consultation between the Forestry Commission and the North East Forest Alliance that followed the storming of Forestry Commission headquarters. In many ways, the people of north-east New South Wales lost their voice, though some would say it was lost some time ago, and that it is disempowerment and frustration that lead to such actions.

Both militant and non-militant actions, whether media-focused or not, have increased in number as the capacity of mass mobilisations such as street marches to deliver large numbers of people has decreased over the past decade. As with the campaigns of the late 1970s and early 1980s, the more militant direct actions remain highly visible at the turn of the millennium, but are actually isolated events. Many of these supposedly more violent actions (as depicted by media reports) are not violent towards people at all, but merely depict a more active form of resistance. For example, in August 1998, three anti-Jabiluka activists, protesting against the expansion of the uranium mine in the Northern Territory, were arrested in North Terrace, Adelaide, during a 'nondescript scuffle' between police and protesters, as police advanced to apprehend protesters writing anti-Jabiluka mine slogans on a building-site wall. Although the protesters more actively defended themselves from the police advance, their practices do not necessarily contravene the principles of non-violent action.

One key outcome of such tactics has been the increasing debate about changing interpretations of the practice. In the earlier forest campaigns, non-violent action was implemented along lines that participants termed 'purist', 'orthodox', 'traditional' and 'strict'. It has also been referred to as the 'Burrowes' interpretation, named after Australian Rainforest Action Group activist Rob Burrowes. Burrowes not only rejects militant direct action, but argues that many protest activities are counter-productive to the Earth. For example, he believes that planning actions in secrecy and 'failing to respect any individual's rights and opinions will only perpetuate malignant global structures' (quoted in Chandler 1992). The example of Rupert Russell 'in perfect repose' while being arrested during a Daintree blockade is an excellent example of this earlier more 'traditional' interpretation.

The more recent interpretation of NVA, based on the term 'active resistance', is still non-violent, but more animated and imaginative in its choice of techniques, and people have the right to defend themselves. As in the first Daintree blockade, active resistance includes those actions that are genuine last-ditch attempts to halt further environmental degradation, such as moving survey pegs, 'sitting protesters in the path of machinery atop giant tripods, and running away from police instead of surrendering' (Chandler 1992). These sorts of actions were implemented at the Chaelundi protests in north-east New South Wales. Earth First! advocate and founding member of the Rainforest Action Group in Australia, Charley Daniel, described the rationale behind these protests in the rainforests of Chaelundi: 'We've been so caught up in the ideology of NVA, but it showed we can be basically non-violent but do more radical things ...' (quoted in Chandler 1992).

Although again dominated by non-violent action and consensual decision-making, some more experienced individuals took part in 'arrestable' active resistance. For example, some participated in what has been called 'black wallabying'. These protesters black out their faces and use camouflage to enter the logging coupes (the current area of logging activity). Once inside the coupe, the protesters identify themselves and logging activities must cease immediately for safety reasons. Where once the more fundamentalist NVA protesters of Daintree would have surrendered 'in perfect repose', the 'black wallabies' play hide and seek. As soon as logging operations recommence, the protesters reveal their whereabouts. Again, logging must cease until they are apprehended. Four were arrested in this manner in the first week of this four-month campaign (Cant 1993, p 2).

Direct action in Australia has been largely characterised by the 'traditional', 'orthodox' interpretation of non-violent action. Militant direct actions have always been in evidence, but are more isolated events. With increasing frustration within the movement, these militant direct actions may have increased in the 1990s, but much of this tension has forged a more creative, less frustrating interpretation of NVA known as active resistance.

In the USA, the major conflict within direct action networks has not been between different interpretations of NVA, but between two factions of Earth First!: one advocating non-violence and the other not categorically rejecting it. The 'wilders' (those not rejecting violence) are basically biocentric wilderness activists who believe that other issues such as 'social justice, anti-imperialism or workers rights' can 'alienate many potential wilderness sympathizers' (Taylor 1991, p 263). 'Holies', on the other hand, although also biocentric, are more humanistic in their focus, and emphasise that social justice must be achieved, and society must be overturned, if ecological integrity is to be retained. Holies argue that activism based on the separation of ecological and social issues will ultimately fail because industrial society itself, not only commercial incursions into biologically rich wilderness areas, destroys biodiversity (Taylor 1991, p 263).

The strategies of these factions reflect their diverse perspectives. Wilders tend to use more militant activities (sometimes violent), while holies generally pursue a less violent path. In Australia, thus far, there has been no major schism within Earth First! Earth First! activists usually advocate more militant actions, although there are notable exceptions. John Seed, an internationally renowned Earth First! activist, does not personally advocate ecotage. He states: 'Destruction of property at this time, in this society, is counter-productive. It will be misconstrued as violence and authorities will act violently against the perpetrators—and against greenies who may be mistaken for the perpetrators' (Seed, quoted in Chandler 1992).

In Australia, however, those who advocate strict NVA usually operate outside Earth First!, which remains an extremely small network. In the Australian environment movement the debate between NVA and MDA tactics has been diffused somewhat as active resistance, a more flexible interpretation of NVA, has been added to the activist's range of options.

The direct action phase of many longer term campaigns, such as the wet tropics, is relatively short-lived. Different networks dominate the campaign at different times. Informal politics is not restricted to one form or another of direct action. Informal networks, with all their power relationships, weave their way through every part of the movement's web. The environment movement symbolically rallies people with diverse goals. Chapter 4 illustrates and celebrates this diversity of ideologies and variety of opinions, strategies and goals.

NOTES

1   In an interview, Nan Nicholson, the landowner and protester who provided a camping area for protesters on her property at Terania Creek, told me that she believed in the effectiveness of the more militant actions, but personally advocated an NVA position.

2   Although environmentalists use Aboriginal arguments on occasions to further legitimise their claims, there is still an enormous gap between the objectives of Australia's indigenous peoples and what is widely considered the white, wilderness movement. Some organisations in the late 1990s—such as the ACF and ANTaR— are working towards genuinely reconciling the two movements.

3   According to many interviews I conducted with people who lived in the area, advice from organisational environmental activists (when they did arrive) was largely ignored. Among these interviewees was Hans Niewenhasen, whose property the protesters lived on.

4   The fight for the Franklin River took place in south-west Tasmania in 1983. Again NVA tactics were employed by protesters in the temperate rainforest. The Franklin campaign was so powerful that it has been credited with changing the national government in 1983. Eventually a decision to 'save' the Franklin was taken by the High Court — the utmost adversarial body in Australia—in a case fought by the newly elected federal Labor government against the conservative state government of Tasmania. We look briefly at the campaign to save the Franklin in chapter 8.

4

THE IDEOLOGY OF THE
ENVIRONMENT MOVEMENT:
THE INVOCATION OF NATURE
AS A POLITICAL METAPHOR

The symbol 'environmentalism' has become a major site of political and social contestation. It is a green prism through which the torch of so many political traditions is now being shone.[1] These include the political philosophies of the conservatives, the liberals and the libertarians, those of the anarchists and the socialists, those of the feminists and many others.[2] The light from these torches is bent, mutated and diffused by its contact with the green prism. Ideas emerge under new names: free market environmentalism, sustainable development, eco-fascism, eco-anarchism, eco-socialism, eco-feminism, social ecology, human ecology, deep ecology, shallow ecology and so on. Some would argue that the new names are little more than old ideas masquerading under new nomenclature, with nothing much having changed.

Others maintain that there is some real gain from what occurs at this site of contestation. This gain does not have to be interpreted within the modernist project of Enlightenment. It does not have to be about the pursuit of progress towards perfection and ultimate truths. Instead, the improvement of ideas may simply lead to their increased applicability to our times: appropriate working truths that define and resolve environmental conflicts.

In this vein, the older ideas are reinterpreted to become more pertinent to an ever-changing world. As they are honed, parts are discarded, parts are improved. Now, proponents of ideas who have long been polarised in their understanding are brought together by the need to synthesise new ideas capable of overcoming vast and varied environmental problems. Perhaps this is a way of cobbling together chains of ideas, inventing different combinations of 'ideas DNA'; or perhaps some new elements are discovered, some genuinely new ideas.

Different cultures define symbols differently. Cultural definitions and perceptions (which inevitably lead to and are shaped by actions) also change over time. These changes are informed by what has preceded them, as well as what is imagined for the future. 'Culture' here refers to that social collective experience that occurs on global, national or local levels, or in even domestic or workplace subcultures. All cultures provide idea clusters, mythologies, symbols and metaphors that offer members identification with a particular grouping, as well as constituting a very real expression of their lives.

'Environment' is such a powerful symbol that it exists in many guises in various cultures. Not only is it defined differently, it is interpreted and acted upon in various ways. There is no one or 'true' environmentalism. This is not to say that some versions have not been dominant, or that some environmental ideologies are not preferable to others. In central European countries,[3] more radical forms of environmentalism can often be characterised as human or social ecology.[4] Their adherents have been largely concerned with the intersection between humanity and the environment. For example, they argue that humanity is part of nature, not distinct from it. The original West German Green Party's 'four green pillars' are peace and non-violence, more representative and participatory forms of democracy, social equity and ecology. But the German Greens see as equally (if not more) important the interrelationships between people. Important issues have emerged that relate to human health and social equity, though advocating an ecological vision of the interconnectedness between humans and the non-human world. This view remains unashamedly anthropocentric (human-centred), though with a deep respect for the non-human domain. In Sweden, for example, a major set of issues that drives much environmental discourse relates to the workplace environment.

In certain parts of the developing world or the 'South',[5] environmentalism has led to different interpretations. Many dissenting environmentalists in the South seek to gain control of their own resources from both Northern and Southern elites. They do not see 'environmentalism' as a luxury, post-materialist ideology of the middle class, but as central to daily survival. Securing appropriate sewerage and solid-waste systems, seeking uncontaminated food and water, and alleviating the associated problems of abject poverty are the issues defining environmentalism in parts of the South. In this manner, the less affluent, majority-world interpretation of the green symbol can often be referred to as an ecosocialist one.

In Australia and North America a powerful interpretation of environmentalism has been a more preservationist approach, exclusive of human welfare.[6] Large tracts of wilderness, national and state parks and reserves have resulted from these ideas. At first glance, protecting non-human areas appears less anthropocentric than the dominant movements in parts of Europe and the majority world. Certain European human ecological traditions are anthropocentric in that they ultimately favour human concerns. From another angle, however, they are less dualistic in their 'natural overview', as they are fully cognisant of the fact that humanity is just one part of nature (however favoured). The resource-preservationist and deep-ecological approaches of the 'New World' are far more separatist, often unconsciously arguing for a clear demarcation between the human world and the non-human world, preserving and conserving areas that are relatively free from human activities and interactions, reaffirming the inherent separation between culture and nature.

In many parts of the world environment movements have contested inequalities and differences in power based on gender, class, race and species. And then there are the organisational cultures, which cross national and ecosystem boundaries. The environment symbol has been interpreted very differently, for example, in bureaucratic and corporate cultures than in cultures dominated by more radical, informal politics. Most bureaucratic perceptions of the environment can be defined as the 'resource conservation and development' approach. This is primarily concerned with 'balancing' development and environment interests. It is fiercely anthropocentric, recognising some limits to material growth, but still seeing nature as 'out there', something that can ultimately be understood and controlled. Another powerful and widespread interpretation emerging particularly from corporate cultures since the mid-1980s has been 'global ecology'. Corporate definitions of culture, in particular, are increasingly enjoying transnational currency, challenging the dominance of differing green traditions that have surfaced in different countries over the past generation. The establishment of 'global ecology', for example, has its origins in global market cultures, defining and controlling environmental concern in its own way: as a series of results due to natural forces, rather than relating to problems created in part by human and resource inequalities. These theories are taken up again in chapters 9 and 11.

ENVIRONMENT AS A SITE FOR CONTESTATION

Three types of arguments can be used to comprehend the current power of the green symbol. First, 'environment' is an associated symbol of 'big N Nature'. It is an all-encompassing banner relating to human nature and the nature of all else in the cosmos. A second argument relates to scientific necessity, making the point that all existence 'depends completely on the continuing availability of inputs such as air, water, foodstuffs, and other resources from the natural environment. Without ecology there can be no economy and no society' (Doyle and Walker 1996, p 1). Even from an anthropocentric, utilitarian angle the welfare of the environment is simply central to survival. David Wells (1993, p 525) writes:

> While there may be immediate and short-term benefits to be gained for particular individuals by polluting waterways, pouring poisons into the air, or destroying tracts of forest, if these activities are carried on for long periods of time, and on a wide scale, then they ultimately must affect us all in highly negative ways. To continue such activities, to be indifferent to the environment in that sense, is simply unreasonable from any human viewpoint ... The environment movement does not really need any more than this argument to make its case ...

This argument recognises a 'natural reality' beyond what postmodernists frequently refer to as the domain of language and discourse. These more essentialist ecological arguments suggest that the environment is a major site of contestation simply because of the centrality of ecology to life, that its dominance in discourses has occurred through necessity. This raises many questions of epistemology (the construction of knowledge) and ontology (the nature of being).

The essentialist viewpoint argues that there are real environmental limits. Regardless of social construction, non-human nature shapes, designs and has a profound impact on human and non-human lives. This argument is also embedded in the concept of scientific objectification: that there are 'natural laws' that provide us with eternal, unequivocal truths about existence itself. This perspective sees the environment movement emerging, in all its disparate forms, because it had to: to a large extent history is environmentally determined.

Ecologists, then, see their science as the study of the interrelationships between different species. They seek to cross the disciplinary boundaries of 'normal science', and to integrate knowledge systems in a bid more fully to understand the natural equations. Once all the inputs are known, then humanity can better 'manage' non-human nature. Ecology's key principles, such as holism, integrity, diversity and interconnectedness, are portrayed as eternal, natural, scientifically proven truths. The secret to averting environmental crises, according to these ecological essentialists, lies with people learning, accepting and acting on these self-evident 'truths' and 'facts'.

A further explanation of why the symbol 'environment' is so prevalent is that it is purely a social phenomenon. Through various subtle plays in history, the green symbol is just a 'place' or 'site' where power is contested. This may have occurred for many reasons. Perhaps the political 'success' of the environment movement has dragged it onto centre stage. History is far more relative in this view. Its advocates argue that the so-called laws of science do not exclusively dwell in a non-human reality, 'out there somewhere', but are, more often than not, created by cultures and societies to justify certain kinds of behaviour, to endorse the power of certain groups over others. This social constructionist depiction of nature perceives different sections of humanity as valuing and promoting different characteristics, and projecting these traits onto non-human nature.

For example, social constructionists may argue that 'ecology' and 'environment' are used as metaphors to justify certain social outcomes, projected onto an imaginary canvas of nature 'out there'. These outcomes portray certain types of politics as the 'natural' and therefore the better way, the true way. In most views of environment—as in the case of nature itself—this symbol is seen as innately right.

There is no doubt that environmental determinism has been used to justify an unequal distribution of power for a number of other purposes. And it is impossible to ignore the notion that there are limits to human and non-human agency on occasions. Yet alarm bells start ringing when metaphors associated with 'nature' and 'environment' are used to severely limit access to power by certain sectors of the community. For example, the phrase 'it's just human nature' has been used to justify appalling behaviour, including extreme selfishness, brutality and wars. This is when 'environment' and 'nature' cease to be ambiguous symbols and emerge as profoundly dangerous ones. All notions of what is 'true human nature' and what is 'natural' are ultimately leaps of faith made by the proponent of the belief. The assumptions are largely ideological and are embedded in the often inexplicit realm of mythology.

THE MANY FACES OF GREEN ECONOMICS: WILL THE REAL NATURE PLEASE STAND UP?

As much of the movement in Australia has traditionally been dominated by wilderness and anti-nuclear issues, the true diversity of ideas within the movement is not often apparent to those outside it. In fact, as we have seen, even within the environment movement some hold the notion that all environmentalists are 'in it' for 'the same thing'.

As there is no one environmentalism, there is no one green economics. Different economic frameworks and solutions are built on different political theories and systems. Many of these systems depict 'natural processes' in different ways. Many of these frameworks and systems use elements of environmental determinism to justify their political and economic paths as 'right' and 'true'. Advocates of disparate theories have developed their own forms of green economics.

GREEN ECONOMICS AND THE RIGHT

Conservatives, liberals and radical libertarians (or neo-liberals) are all part of the political right. They usually argue that green economic solutions should be sought within the frameworks of both advanced industrialism and capitalism (politics is subservient here to economics). Interestingly, each ideology offers its own view on what 'human nature' is. Conservatives argue that humanity is innately evil. A wonderful example of the conservative political philosophy is found in William Golding's *Lord of the Flies* (1955). In this famous book a group of boys are stranded on a deserted island without the recognised authority of adults to guide them. The boys resort to 'the rule of the beast', which lies at the heart of our 'human nature', and develop a society based on greed, fear, coercion and force. One of the morals of the story is the necessity for authority imposed upon citizens 'from above'. It is not surprising, then, that conservatives would support monarchies and aristocracies, for example, arguing that 'ordinary people' are limited in their capacity to solve their own problems. Indeed, conservatives argue that there is a 'natural order' that allocates certain privileges and powers to certain castes within societies. This ensures a smoother running of a given society than if its operation were determined by such forms as democracy.

Conservative political theorists see 'natural history' as something intrinsically violent: a universe at war. How then has conservatism intertwined with environmentalism? Robyn Eckersley (1992, p 21) comments:

> ... there are some notable points of commonality between conservatism and many strands of environmentalism. The most significant of these are an emphasis on prudence or caution in innovation (especially with respect to technology), the desire to conserve things (old buildings, nature reserves, endangered values) to maintain continuity with the past, the use of organic political metaphors, and the rejection of totalitarianism.

One of the most obvious instances of the conservative green economic response relates to 'scarcity economics'. The earliest proponent of this theory was Thomas Malthus, 'the dismal parson'. Malthus (1826) presented his 'population principle' 200 years ago, arguing that the Earth's resources would increase arithmetically, while the Earth's population would grow geometrically: this meant that the Earth's resources were finite, and could not support an ever-expanding population. This line of reasoning has been used

> to justify birth control programmes and to urge no response to poor people in need as a result of famine and drought. It was an extremely popular argument in the 1960s and 1970s, championed by both Garrett

In recent years green
and black movements
have found common
ground in wilderness
and anti-nuclear issues.
Phil Bradley

Hardin and Paul Ehrlich. Their argument begins with the assumption
that there are already too many people consuming the Earth's limited
resources ... A 'population bomb' is set to explode ... With such a narrative other questions emerge: what is the Earth's 'carrying capacity', and
who should have access to increasingly scarce resources? (Doyle and
McEachern 1998, p 71)

The debate about 'carrying capacity' within green movements is an
essentially conservative one. It assumes 'natural limits' placed on earth-
ly existence by a transcendental moral authority, 'Nature' with a capital 'N'. The conservative green economic response has largely been
about reducing absolute numbers of human beings, usually from the
developing world. Poverty is seen as one of the main causes of environmental degradation, not a result of it. Australians for an
Ecologically Sustainable Population is one conservative, green organisation offering economic solutions along these lines, opposing further

immigration on the basis of ecological limits. Elements of this conservative response appear in the recently emerging ideology of 'deep ecology', though other deep-ecological tenets are derived from other traditional political philosophies, such as anarchism.

The liberal tradition, though still from the right, sees natural processes, including human nature, in a substantially more generous light. Classical liberal theorists such as John Locke and Jeremy Bentham argued that the vast majority of people are good but are sometimes hampered by an anti-social minority. So nature, in this view, is not in a perpetual state of war, but can get into minor skirmishes from time to time. In a bid to protect the majority from the anti-social minority, liberal theorists argue for the establishment of a moderate state, not an authoritarian Leviathan,[7] to promote the interests of the majority. Although liberalism does have some notion of 'greater good', it still emphasises the rights of the individual, supporting the idea of autonomy for each citizen within responsible bounds regulated by governments. Liberalism is heavily reliant on private property and is closely associated with capitalism and its emphasis on continuing economic growth. It assumes that all citizens have equal access to power and resources, often denying inequities based on class, gender, race and species.

The liberal green economic position is sometimes called 'shallow ecology'. It has led to legislation that provides environmental protection while not unduly interfering with the everyday mechanics of capitalism. One instructive example of a liberal green economic response was the development of much environmental legislation in the 1970s and of the Environment Protection Authority (EPA). The EPA—at state and federal levels—was established to monitor the excesses of an environmentally degrading minority and to punish excesses through fines, restrictions and further regulations.

Radical libertarianism (or neo-liberalism) is an extreme, simplistic and, currently, virulent strand of liberalism. The radical libertarian/neo-liberal or 'market liberal' position is reminiscent of laissez-faire economic systems that emerged with the industrial revolution in Europe in the nineteenth century. These economic libertarians offer their own interpretation of Adam Smith's proposition that an 'invisible hand' operates within capitalism whereby the pursuit of individual self-interest leads, unwittingly, to the advancement of the common good. They therefore seek maximum autonomy for the individual, and regard any government interference as a direct threat to this freedom and autonomy. For radical libertarians, the 'state of nature' is the survival of the fittest individual leading to the survival of the fittest collective, though the condition of the collective is hardly important. This is remarkably similar to popular and often misguided interpretations of Charles Darwin's theories of 'natural selection' and Herbert

Spencer's theories of social evolution. It is hardly surprising that both emerged on the coat-tails of the industrial revolution and the initial surge of capitalism. This most extreme version of liberalism is really an anti-society ideology. Society is nothing more than a collective of atomised individuals; it is not designed to promote infrastructures of welfare, health and education, let alone protect and uphold a collective sense of 'environmental good'. In the tradition of laissez-faire, the poor and the sick simply deserve to be so, and the 'management' of nature is best left to market forces, as the market *is* natural. This right-wing tradition now dominates Australian federal politics under the conservative coalition.

Green radical libertarian approaches have also been renamed free-market environmentalism. Attempts to make the national accounts 'green' and promote 'green' consumerism provide excellent examples. In the first instance, green free-marketeers argue that a key problem is that not all of nature has been allocated economic value. Eckersley (1993, p 29) comments:

> Many natural resources are regarded as gifts of nature with a zero supply price, and the accumulation of wastes is regarded as a 'negative external-ity'. For example, our woodchips exports are registered as national income but no allowance is made for 'natural capital depreciation' (i.e. the depletion of our forests).

So, rather than arguing that some natural attributes are beyond monetary value, green radical libertarians seek to include all of nature into their market strategies in a bid to resolve environmental problems while still pursuing economic growth under capitalism (see Eckersley 1993). Interestingly, while libertarians insist on the user-pays principle elsewhere, they usually reject the polluter-pays principle. Green consumerism is another central plank of all liberal green responses, but it is most avidly championed by the radical libertarians.

Imagine you are an alien being, having just arrived from some far galaxy, and you observe a civilisation that accords value to all things on the basis of how readily they can be consumed. It may appear quite bizarre, but this is exactly how value is accorded in liberal economics. Because of its associated commitment to the role of the individual, much emphasis is placed on the environmentally educated consumer. The argument runs: 'Once people start purchasing more environmentally friendly products, then the market will have to respond and produce more of the same.' With an emphasis on informed consumption is the notion that producers will self-regulate. This position is still firm-ly entrenched in the politics of pluralism and capitalism, which assumes that producers exclusively respond to markets rather than influencing them through marketing and advertising.

GREEN ECONOMICS AND THE LEFT

Two traditional political ideologies with origins more closely aligned with the left are socialism and anarchism. Some of the key features of Marxism, which is the forerunner of most forms of socialism, can be described as follows:

> Marx saw the conflict between classes as the 'motor of history'. Inevitably, capitalist development would produce 'wealth at one pole, misery at the other', and an increasingly polarised society. It was Marx's expectation that sharp class antagonism would provide the basis for radical political change when the proletariat united to overthrow the social, economic and political power of the bourgeoisie and usher in a new stage in world history, socialism followed by communism. (Doyle and McEachern 1998, pp 44–7)

Marx argued that once capitalism was overthrown, the Earth's resources could be adequately managed in a planned, centralised economy that represented the interests of the vast majority of the people, not just the wealthy 'bourgeoisie'. Indeed, the health of the society as a whole becomes a paramount goal under socialism.

Under socialism, 'nature' can be seen as a community, working for the ultimate survival of the whole ecosystem. The ecological concept of 'holism' nicely matches with this political thought. How does this political philosophy generate its own brand of green economic responses? In a more centralised green economy, the state would not only monitor and legislate against the excesses of environment-degrading companies, but would pursue a green economy by being the major shareholder in it. The issue here is not whether there are inadequate resources to go around, but that under capitalism these resources have been wrongly concentrated in the hands of the few. Under socialism, the people, as a collective, would own their resources. Consequently, green socialists are not concerned with population 'carrying capacity' or 'resources scarcity': they see poverty as one result of capitalism, not the fault of the poor producing too many children.

A green or eco-socialist economy would seek to reduce overconsumption in the Earth's most affluent classes and societies, and it would seek to redress the unequal distribution of resources. Green socialists are often criticised as being unwilling to accept the fact of ultimate biophysical limits. As Marxism is primarily a critique of capitalism and demands a complete societal response and revolutionary structural change to the economy, it is not surprising that practical green socialist economic responses in the Australian context are virtually non-existent. Eckersely (1992, p 31) comments:

> ... ecosocialists argue that we must move toward a situation where the state is no longer fiscally parasitic on private capital accumulation. They

argue that the contradictions of the market must, for the most part, be eliminated (by moving toward a democratically planned economy) rather than simply 'managed'. However, while the analysis may be sound, ecosocialists have yet to develop a satisfactory alternative prescription.

Although Eckersley is correct in the context of the industrial North, this comment is not particularly apposite to the South, where eco-socialist prescription is quite simple: to wrest ownership of environmental resources from the wealthy elites of the North and South, and deliver them to the people.

Whereas conservatives are pessimistic in their 'view of nature', and liberals are rather more hopeful, anarchists are the eternal optimists. Anarchist theorists see nature as an inherently beautiful place. They argue that in its original state, nature is innately good, as are its people. Authority is not developed through necessity to protect people from themselves, but rather as a means of the powerful to promote their own interests at the expense of the interests of others. Anarchist theorists such as Murray Bookchin (1980) view a healthy society and a healthy nature as one, as do most political philosophies, but anarchists present the relationship between the component parts and the whole differently. Bookchin sees a healthy ecosystem (and society) as diversified, interconnected, balanced and harmonious. An anarchist society must be non-hierarchical, decentralised and self-governing.

Although they share the socialists' interest in overconsumption and unequal distribution of resources, anarchists do not support the centralist tendencies of some socialist positions. Instead, anarchists argue for maximum local autonomy, decentralisation and a lack of formalisation in decision-making. Local Exchange Trading Systems (LETS) is a non-profit organisation with an emphasis on decentralisation that emerged in Canada in 1983. It arrived in Australia in 1986 and now there are over 250 LETS systems operating in Australia, providing local currency and barter schemes. LETS Association Inc. describes their operation as follows:

> LETS members are a collection of people who have all agreed to buy and sell with the group in return for local currency. When a transaction is to take place, the people involved negotiate the value of the deal ... The member who receives the goods or services authorises units to be transferred from his/her account to the account of the supplying member ... Local currency is created by the community and can only be used in that local area. It reflects the real wealth—the people's initiative, skills and efforts. (LETS South Australia 1997)

Bioregionalism is another result of the green anarchist response. Bioregions are naturally occurring boundaries that define ecological and social communities. The Murray–Darling Basin scheme, which crosses the political boundaries of three states, is an excellent example of a bioregional management system adopted by state and federal governments. Bioregionalists also emphasise the importance of place. The concept of place sees individuals as having a physical and spiritual attachment to a community of other people and to a biophysical, defined space. They reject the new-found extreme mobility of advanced industrialised societies. Thus green anarchist economics rejects the globalisation of economic systems with its attendant reliance on free markets and mobile capital.[8] Introspective groups such as Earth First!, the Nomadic Action Group, the EcoAnarchistAbsurdist-AdelaideCell?! and the Kumarangk Coalition are further examples of green anarchist or social ecological politics being played out in Australia.

The 'environment' symbol and all its associated green metaphors have become major places of political and social contestation and will assume even greater importance. The symbol's dominance can be explained by its close semantic relationship with nature itself (and the ease with which it has been adapted to suit a wide range of political purposes), and the very real and significant environmental limits on our lives. Finally, the political durability of the environment movement has fostered societal acceptance of the importance of the symbol.

Its numerous interpretations derive from both right and left, yet— to borrow an anecdote from the German Greens—green is neither left nor right, but in front. It crosses a range of spectrums that defy the limited, traditional left–right conception of politics. 'Green' ideas sit in continuums as diverse as those of patriarchy–matriarchy; formality–informality; 'Southern'–'Northern' interpretations; and the principally pluralist, apolitical view of wilderness preservation at one extreme and the intensely structuralist stances of the anti-nuclear activists at the other. Movement participants need to understand and respect this diversity of ideas and mythologies if they want to be part of an active, vibrant social movement.

NOTES

1   The first time I heard the term 'prism' used in this way was by Peter Hay at the Centre for Environmental Studies at the University of Tasmania in September 1987.
2   The interactions between feminist and green political theories are astoundingly rich and diverse. They demand special attention that cannot be sufficiently provided here. I have introduced some of their variations elsewhere (Doyle and McEachern 1998), but for an excellent overview of differing 'ecofeminist' positions, see C. Merchant, *Radical Ecology* (1993).
3   There have been many manifestations of 'environmentalism' in Europe's many different national cultures. To talk of Europe, or parts of Europe, simply serves as a useful comparative generalisation.
4   Some writers go to some length to separate human from social ecologists. There

are some useful determining characteristics. Human ecologists usually argue from liberal positions, advocating change from within established systems, while social ecologists, with closer connections to anarchism, are more revolutionary in the processes they advocate for change. The former are more atomistic and individualistic in their depiction of natural society, while the latter are highly ecological. They share the belief that humanity is part of nature and, at the same time, its most important part.

5   I use the terms 'developing', 'newly industrialised', 'less affluent', 'majority world' and the nations of the 'South' to signify those cultures that Australians traditionally saw as part of the Third World.

6   This is often confused with the ecocentric approach. The two share an interest in protecting the human world. Preservationists do it on the basis of human values (see Fox 1990), such as a symbolic monument, or recreational value, whereas ecocentrists (or deep ecologists) see the environment/nature as having value in itself, independent of humanity's perceptions. Also, ecocentrism is the least anthropocentric of environmental ideologies, seeing humanity as just one part of nature, equal, not superior, to other life forms.

7   Leviathan is a term used most famously by Thomas Hobbes, who argued for a 'rule of iron' by a machine-like monarch. This was the only way the Earth could avoid catastrophe. Neo-Hobbsians have emerged to advocate similar positions. For a thorough analysis of the neo-Hobbsian position, see Walker 1988.

8   The noted American poet Gary Snyder has done much to further the cause of bioregionalism. See, for example, Dardick 1986.

5

ENVIRONMENTAL
NON-GOVERNMENTAL
ORGANISATIONS

> Non-Government Organisations (NGOs) are the most visible players in environmental politics around the globe. They are involved in many different spheres of politics, from the local community level, through the politics of the nation-state, to international politics. They exist in both the predominantly non-institutional domain of social movement politics and in the institutionalised milieu of political parties, administrative systems, governments and beyond. (Doyle and McEachern 1998, p 81)

Most people would perceive the environment movement simply as a cluster of environment organisations. This is largely because pluralist perspectives of politics dominate, and there is little recognition of the informal realm. NGOs are only one political form, one part of the environment movement. Nevertheless, they have been an extremely visible and important force shaping green policy in this country for a generation.

Formal environmental organisations—or NGOs—have a structure that allows them to play politics in specific ways. They are based on constitutionalised relationships, with many larger organisations employing professionals to service the daily needs of the organisations and to perform bureaucratic tasks. Green NGOs are often caught in two worlds: both the institutional and non-institutional domains. Some writers refer to these NGOs as voluntary associations or the third sector: that area of political relationships that falls between the private and public sectors, as well as including characteristics derived from these two more powerful sectors (Etzioni 1968).

CHIEF CHARACTERISTICS OF FORMAL ORGANISATIONS

Environmental NGOs are political organisations. An organisation, like a social movement, has distinctive characteristics.

CONSTITUTIONALISM

Probably the most obvious feature of the organisation is the official statement of its goals: the constitution. This is the main feature that distinguishes formal political activity from the informal. The constitution is symbolic of the acceptance of legitimacy and rationality, it publicly states the organisation's intention to work within the power structures of the state. Constitutions also establish certain rules that set out how power will be dispersed throughout an organisation, how decisions will be made, and how people will relate to each other.

Although they are diverse, environmental NGOs are far less heterogeneous than the range of groups found within environment movements, as certain rules for playing established politics are shared. Sometimes, of course, these rules are open to interpretation.

NGOs' constitutional structure makes it relatively easy to study them, since it leads to the production of formal records. In turn, this more official historical documentation is supplemented by media reporting, and then a whole range of secondary commentary. In part two of this book, more emphasis is placed on this more formal organisational realm, largely because material is available, rather than because of the perception that these green NGOs have an inherent centrality in movement-wide policy-making processes.

COMPLEXITY

> As civilization has increased in complexity, we have become more involved in large-scale organizations. Many commentators on modern social life have observed wide-spread organizational involvement as a basic and pervasive characteristic of our current society. Modern man is man in organizations. (Brinkehoff and Kunz 1972, p xi)

The above quotation opens Brinkerhoff and Kunz's work on modern organisations and their environments. Complexity, it is regularly argued, is an inherent characteristic of modern social organisation. There are other factors, however, that could account for this view. One relates to scientific tradition. Complexity has been used over the past few centuries by scientists as a criterion for analysing animal, plant, and ultimately human organisms. Darwinists used these concepts of complexity to differentiate between 'lower' and 'higher' forms of life on the unilinear path of progression towards ultimate evolutionary perfection.

Among other early social observers, Saint-Simon and Comte made reference to a future human society being run by a complex system of organisation based on scientific 'knowledge' (Gouldner 1959, pp 400–10). Herbert Spencer coupled these ideas with notions pertaining to 'the laws of nature'—which would later become more popularly recognisable as Darwinian—in his bid to analyse human societies. He compared primitive societies to 'simplistic' life systems in primitive organisms. If the social organisation of a given society became more complex, it also became more 'progressive' on the evolutionary scale. In a normative sense, complexity was 'good'.

Many present-day social scientists are still affected by these ideas; they tend to assume that the contemporary organisation, due to its modern nature, is the most complex and progressive form of human relationship. Their one-way view of history binds them to the idea that most human collective forms are continually evolving towards increased formalisation and complexity—the Weberian concepts of increased bureaucratisation we touched upon in chapter 2 are excellent examples. This type of research into human interaction assumes prior to the actual study of these relationships that they are complex, perhaps creating complexity where none exists.

The study of organisational objectives has taken precedence over all other forms of goal-oriented human integration in capitalist society. Within this scenario, organisations have been rationalised, defined, redefined and then rerationalised so many times that they have often taken on complex characteristics because of this emphasis, rather than because they are inherently complex. The widely held view of the modern organisation is that it constitutes a more developed, ordered and rationalised social grouping. As categories are created and, in turn, divide and subdivide, certain clusters of organisational members are differentiated from others. This results in the increased complexity of the organisational apparatus, and organisational goals begin to share similar characteristics.

Also, complexity may dominate discussions concerning organisations because of the way in which Western society wishes to perceive itself. The West has used complexity for centuries to classify the world. Texts use this criterion with little explanation. For example, Claessen and Skalnik's huge work *The Early State* (1978) begins with four chapters on the theory of state origins. In this discussion, they often refer to classification of 'early states' based on degrees of complexity, but they do not investigate the intellectual origins of the selection of complexity as a criterion. It is a characteristic that this dominant society holds most dear in itself. Yet it is a relatively arbitrary characteristic chosen by a dominant society to reinforce that dominance. Thus, the organisation, the symbolic fulfilment of social dominance has been termed the most 'complex' and progressive social entity.

Another reason why the formal organisation is widely perceived as complex relates to its structure. When a collection of people become an organisation, they are implicitly demanding to have input into the established system, but at the same time, they are recognising the norms, values and expectations within that system. Political science and sociology have concentrated their efforts on the study of organisations, as the study of legitimate political groupings offers legitimacy and respect for those who observe and comment on them.

COVERTNESS

One other characteristic of the constitution deserves mention here. Its official front is quite often different from the goals of the individuals who make up the organisation. For this reason it cannot be used to portray organisational objectives in real terms, although this practice is very common in political science. In the more informal environment network and group, goals are most often overt. An organisation that operates in the environment movement, however, is far more covert about its goals. Sometimes goals are purposely hidden from public view to protect the organisation's members or employees. The cover of constitutional goals may also be used for a number of practical reasons. For example, on occasions, it may enable organisations to attract more funding or support than they would otherwise (Pfeffer 1981).

CONSERVATISM

The goals of the environmental organisation are often more conservative than those found in networks and those groups determined to maintain their informality. In forming an organisation, individuals accept the established political game as being the only 'legitimate' way of achieving their goals. The environmental organisation must accommodate the structural constraints of the established order. In turn, accommodation often leads to the adoption and sharing of certain goal characteristics that manifest themselves in structural ways. The organisation may become hierarchical. It may emphasise the antithesis of the more informal political forms described in earlier chapters. These may include: vertical role differentiation, the development of subordinate relationships with other organisations, autocratic choice of goals and strategies, and increasing emulation of the operational style of the dominant order—such as the state and corporations—in its bid to become more 'effective' and 'successful'.

The adoption of organisational structure does not allow the pursuit of more radical environmental goals (as pursued by some networks and groups) and openly endorses the role of political reformism in our society. Bill Devall (1981, p 319) writes from a deep ecological perspective that the environmental organisation is largely an instrumentality of reformist environmentalism or 'shallow ecology': 'While reformist

environmentalism is content to work with the bureaucracies and legislatures of modern societies such as the Sierra Club, Audubon, etc.,
deep ecologists harbor a deep distrust of big organizations.'

THE DISTINCTION BETWEEN PRIMARY AND SECONDARY ORGANISATIONS

In their numerous sociological works on the environment movement,
Faich and Gale divide organisations into two types: expressive and
instrumental. They focus on organisations involved, in some manner,
in 'nature recreation'. They argue that there is a distinction between
those organisations primarily pursuing nature recreation objectives
(expressive) and those dominated by political, conservation-oriented
goals (instrumental). They write: 'Expressive organizations would
attract members primarily seeking outdoor recreation activities and
companionship, while more instrumental organizations would draw
members sympathetic with the organization's external action program'
(Faich and Gale 1971, p 275).

Faich and Gale add that there is a transition from expressive organisations to those more politically oriented. Indeed, some recreation-
based organisations in the Australian environment movement do
match this description. For example, the Queensland Speleological
Society was initially set up by people interested in the exploration of
caves. Over the years it became an intensely 'instrumental' organisation
fighting for the preservation of Mt Etna Caves and other issues.

The usefulness of Faich and Gale's distinction between different
types of organisations is limited, for it only includes those organisations
involved in 'nature recreation'. There are many other types of organisations involved at times in the movement that are not recreation-
based. Another more appropriate classification can be used to
differentiate between all movement types based on the distinction
between 'primary' and 'secondary' organisations.

Primary organisations are those constitutionalised groupings that
primarily seek environmental goals. Most environmental NGOs are
primary environmental organisations, ranging from national and international organisations such as ACF, TWS, WWF, FoE and Greenpeace,
to the state-centred 'umbrella organisations', which include the conservation councils and environment centres and numerous local-level
formal organisations. Secondary organisations often undertake environmental goals in a temporary manner, in pursuit of other, non-envi-
ronmental objectives.

Most of the organisations chosen for review in this book are primary environmental organisations. These are the most studied. Many
other organisations, however, while not normally perceived as environmental organisations, temporarily become part of larger networks

inspired by some environmental objective. One example of this is the Phantom Club.

THE PHANTOM CLUB

The primary objective of the Phantom Club is to dedicate the organisation to the Phantom. A letter to members reads:

> The Phantom swears 'The Oath of the Skull' and promises to dedicate his life to the destruction of greed, cruelty and injustice wherever it is found ... He is good, clever, strong and unafraid, and has always fought on the side of the weak against the oppressor, with good against evil ... To this end, The Club provides Members who support the 'Phantom Philosophy' the opportunity to express their feelings by participating in Club activities. In other words, Members are encouraged to bring out the little bit of Phantom that lives in all of us. (Henderson 1985)

The Phantom is a comic-book character who has reached cult status around the world. 'Skull Cave'—Australian headquarters of the club—is in Brisbane. The story of the Phantom begins early in the sixteenth century. While exploring eastern waters, a ship was attacked by the notorious Singh pirates—the only survivor was the captain's son. 'On the skull of his father's murderer, this angry young man swore on oath to fight evil wherever it is found—he was the first Phantom' (Henderson 1985). In May 1986 the Phantom (played by a Brisbane club member) appeared at a rainforest rally held at the Roma St forum in Brisbane. The *Courier-Mail* was quick to spot 'the Ghost Who Walks'. The Phantom told the reporter why he was participating in the rally:

> All creatures need this forest, the animals to live in and man to retreat into and enjoy ... The Phantom wants people to take a long-term view. Both sides have some good points, what we need is to be less selfish, to talk and reach a middle ground ... We must all step out of our own jungle to make the world a better place. (*Courier-Mail*, 16 May 1986)

In an interview with the Phantom (club president John Henderson) it was ascertained that the club became involved in the wet tropics campaign 'to fight against the forces of evil'. 'After all', the Phantom said, 'isn't that what we are all aiming for.' Again, the myth of the common goal comes to the fore. 'Also', the Phantom added, 'I suppose the Phantom lives in the forest, and if you take the forest away, then the Phantom has no place to live.'

More important to the Phantom Club's involvement with the environment movement than 'shared goals' were personal ties linking two organisations. Ken Evans, a former secretary of the Phantom Club, was also an active conservationist with ties to the Queensland Conservation

Council. The choice of the Phantom Club may at first glance seem somewhat ridiculous, but it illustrates the concept of a secondary environmental organisation perfectly. In addition, the Phantom Club, however absurd and bizarre, has been a full member of the major state federation of environment organisations in Queensland, the Queensland Conservation Council (QCC). Table 5.1 uses Charles Perrow's (1970) definition of 'goal types' to classify the different kinds of goals pursued by the Phantom Club.

TABLE 5.1
GOAL CHARACTERISTICS OF THE PHANTOM CLUB

1   Societal goals: To 'fight against evil'.
2   Output goals: comic swaps, phantom paraphernalia, T-shirts, etc.
3   Product goals: 'participation and fellowship rather than aggression and competition'.
4   Systems goals: 'Everything is done with only one goal in mind—to build a bigger and better club'.
5   Secondary or derived goals: To save the wet tropical rainforests.

Source: Based on Henderson 1985.

TWO INTERNATIONAL PRIMARY ENVIRONMENTAL ORGANISATIONS

We now turn our attention to primary organisations. A number of international green non-governmental organisations operate in Australia. Probably the most recognisable of these are Greenpeace, Friends of the Earth (FoE) and World Wide Fund for Nature (WWF). WWF is the most business- and government-friendly environmental NGO. There is some debate as to whether WWF is still actually part of the movement, or whether it has simply become a high-level, wise-use-style front group for industry. Our discussion focuses on Greenpeace and Friends of the Earth.

GREENPEACE

On 10 July 1985 the Greenpeace ship *Rainbow Warrior* was sabotaged by two limpet mines planted by the French secret service. In Auckland Harbour, New Zealand, it lay keeled to one side and was later sunk to form a reef. One crew member, Fernando Pereira, died when the second mine went off. The *Rainbow Warrior* was in the area moving residents from the Rongelap Atoll, which had been contaminated by US nuclear testing in the 1950s, and was to protest against the French nuclear testing at Mururoa Atoll. The action of the French outraged governments and communities in New Zealand, Australia and around the globe. Eventually the French Defence Minister was forced to resign. The sinking of the *Rainbow Warrior* boosted Greenpeace's profile in Australia and New Zealand.

Greenpeace originated in Canada in 1971. It has become the world's largest and most influential environmental organisation with approximately five million supporters worldwide. Its provocative and sensational protests over whaling, tuna fishing, nuclear testing and toxic waste have received worldwide coverage, particularly in the 1980s. The Greenpeace Australia budget swamps all other environmental groups and organisations at nearly $9m (Greenpeace 1996). In contrast to other environmental organisations in Australia, Greenpeace does not accept government grants, so that changes in federal funding arrangements through the 1990s did not affect its income. A gradual decline in income from community sources, however, forced it to announce the closure of all state branches except the one in Sydney in 1996 (ABC 1996a).

At the high point of community and government concern over environmental issues in the late 1980s, Greenpeace Australia was involved in a vast array of environmental issues, with little overall strategy apart from gaining maximum publicity. Greenpeace International's head office in Amsterdam believed that Greenpeace Australia was not taking full advantage of community support for environmental issues. It therefore appointed Steve McAllister, an American Vietnam veteran, as Greenpeace Australia's new chief executive to mould the local organisation into a more 'proactive' one. He said he had been sent to 'clean out' the hippie image of the local outfit and bring it into line with Greenpeace in other parts of the world. He immediately sacked a quarter of the campaigning staff and replaced them with people 'better suited' to the new corporate, conservative and bureaucratic requirements (Notion 1991).[1]

McAllister set up a canvassing division, which roamed every capital city five nights a week. The division was initially very successful, with canvassers selling subscriptions on a 35 per cent commission (Smith 1992). The group's subscriber numbers rose from 6000 in 1988 to 93,000 in 1991, with revenue rising just as dramatically, from $2.3m in 1989 to $8.8m in 1993 (O'Neill 1994). It is important to note that subscribers are not actually Greenpeace members. The payment is essentially subscription to the quarterly magazine *Greenpeace Australia News*, and subscribers receive a regular assortment of Greenpeace marketing publications. There are still only about 50 voting members of Greenpeace in Australia with, new members having to be nominated by an existing member.

The bonanza did not last, with the onset of the recession and some tarnishing of Greenpeace's public image over its sensationalism and media-hungry style—such as the row over Nufarm Chemicals in 1990—and its top-down, hierarchical and centralised decision-making structure. Inevitably, revenues started to fall. In May 1994, in response to decreasing revenues, Greenpeace sacked 30 of its 70 full-time staff, and 100 of its 140 casual workers, disbanding the canvassing division, which was costing about $1.9m in wages annually (O'Neill 1994). The canvassing was described as 'pyramid-selling environmentalism' and had been criticised for using 'hard-sell' techniques (Notion 1991). Despite the disbandment, however, for the year ending 31 December 1995 total fund-raising expenses were still a very significant $2.13m (Greenpeace 1996).

The 'Friends of the Rainbow Warrior Monthly Giving Program' was set up by executive director Paul Gilding and manager of finance and administration Mara Bun in 1993 to bring in more revenue, and was designed to supplement or replace the usual $35 annual subscription fee. The program deducted $10 every month directly from people's bank accounts (totalling $120 per annum) and by May 1994 was contributing $300,000 a month to Greenpeace coffers, which, ironically, allowed Greenpeace the financial security to sack its canvassers. In 1995 the program brought in over $3.2m in revenue while regular annual subscriptions brought in only $1.6m (Greenpeace 1996).

With a big 'win', convincing Shell not to sink its Brent Spar oil platform, 1995 was a profitable year. When new French President Jacques Chirac resumed nuclear testing in the Pacific in the same year, it appeared—temporarily at least—that the good old days of protest were back with Greenpeace sending ships, this time carrying Australian politicians, to the test sites in protest. Greenpeace was back on the nightly news and the funds rolled in. Despite the popularity of this campaign and the boost in revenue from the Monthly Giving Program, revenue began to fall dramatically following the end of nuclear testing.

During this period there were also staff problems caused by the clash of cultures witnessed in other environmental groups and organisations. In August 1995, Greenpeace chief executive Midge Dunn left the job in 'acrimonious' circumstances after only four months when staff refused to work for her. Dunn was a career public servant and her management style was contrary to the culture of the organisation. As one long-time staffer put it, 'people aren't paid a vast amount of money in this organisation and they don't come in here to be treated like kids' (Bita 1995).

The response of Greenpeace was different to that of the other organisations under review. With the appointment of Bronwyn Boekenstein as executive director and the branch closures, Greenpeace dropped any pretence to being a grassroots operation and effectively became an inaccessible branch office of the international group.

Morale was particularly low in the organisation following the financial problems and rationalisation of 1996. This resulted in a much lower profile in the late 1990s.

Greenpeace has always had limited negotiations with governments. Its decision to withdraw from the government-sponsored ESD process, and join TWS in boycotting it, seemed to be vindicated by the deafening silence that greeted the release of the $350,000 study the ACF prepared for the process (Woodford 1995). Unfortunately, Greenpeace has also had extremely limited negotiations with other green NGOs. It rarely coordinates or clears its activities with local NGOs already operating in an area or on a issue. An excellent example was its campaign aimed at preventing a further decline of the southern blue-fin tuna off the Great Australian Bight in late 1997. It didn't contact the Conservation Council of South Australia (the peak group in that state) to discuss strategy. It imagines itself as a power unto itself, and its non-democratic structure is excellent evidence of this.

At one stage Greenpeace portrayed itself as the enemy of business. Many of its protests have been aimed directly at corporations. One particular move in recent years contradicts this image. Its international publication, *Greenpeace Business*, is a clear attempt to involve the business community in Greenpeace activities. The subscription cost of almost $A200 for six issues, each of eight pages, is aimed squarely at the corporate market, with each issue designed for outer office coffee tables throughout the business world.

It is an attempt to make business 'greener' by coopting the values and language of the business community. On the front page of its June–July 1996 issue, under the headline 'Corporate actions may save North Sea fish stocks', Greenpeace 'congratulates Unilever and Sainsbury's for putting the precautionary principle into practice' and clearly sides with business against government:

> As in the Brent Spar episode, the UK Government is being sidelined once again on strategic environmental issues. If any progress is to be made in saving the North Sea ecosystem, it will be the result of an alliance of *corporate self interest* with the vision and determination of environmental organisations. [italics added]

Greenpeace no longer sees business as the environmental bogeyman. It now considers 'corporate self-interest' to be one of the keys to environmental 'sustainability', not an inhibitor. Greenpeace describes this shift to a more corporate image as being more proactive, which is, in itself, terminology derived from corporatespeak. Brow (1997, p 50) writes:

> This shift can be seen in the joint Greenpeace Australia bid for the design of the Olympic Village for the Sydney 2000 Olympic Games. The Games

provided an opportunity for Greenpeace to showcase the latest environmental technologies. The Green Freeze, a fridge which was 1005 CFC and HCFC free, was a major step in promoting solutions.

It is good to see Greenpeace not depicting all corporate players as 'the forces of darkness', a term used too often by environmentalists. Some people working for large companies are committed environmentalists. These new corporate campaigning techniques must be tempered, however, with the understanding that market-based solutions are not enough to combat many forms of environmental degradation. While certain incremental gains can occur due to green technologies produced by markets, it must be remembered that the profit imperative is the raison d'être of the corporate form of human relationships. Even enlightened corporate self-interest cannot replace government responsibility to act on behalf of citizens. The impetus to make business greener also has a negative side if it is not carefully managed. In this new corporate milieu, environmental concerns are more likely to be coopted by business interests, leading to increased profit margins and increasing environmental problems.

FRIENDS OF THE EARTH

Like Greenpeace, Friends of the Earth is an international environmental organisation with a strong Australian following. In contrast to Greenpeace, however, FoE is extremely decentralised and grassroots-oriented. The organisation has more than 10 branches in Australia, and 56 branches worldwide, which are given a large amount of autonomy. FoE branches in various countries have different political styles and structures. The constitution requires at least one annual meeting and actions that comply with the FoE philosophy—in the Australian experience of the organisation—but there is little bureaucracy compared to other international organisations such as Greenpeace, WWF and HSI. It operates as informally as possible within the constraints provided by the constitution.

FoE is essentially non-hierarchical and uses consensus decision-making. There are two annual national meetings where groups send delegates, when possible, and at these meetings there is no fall-back position to the consensus decision. The Fitzroy group, the largest in Australia, does have the fall-back position that if after a second meeting consensus is still not achieved, then a vote is taken. In the experience of long-time FoE activist, Cam Walker, this fall-back position has never been used.

Walker became FoE's first national liaison officer in 1995 when the national meeting decided to adopt a modicum of formalisation at the national level by adopting five national campaigns: forests, sustainable

Australia, sharing the land, uranium, and wetlands. FoE's strong social focus is indicated by its commitment to Aboriginal issues in the sharing the land campaign, and its centrality in the FoE philosophy. The first paragraph in the pamphlet sent to prospective FoE groups states:

> The FOE groups are perhaps best characterised by the fact that they are environmental groups with a strong emphasis on social justice ... The understanding of the FOE network is that you can't separate social and environmental issues—both need to be seen in the context of the society in which they occur, and true environmental sustainability will not be realised without the establishment of socially equitable societies. (FoE undated)

Unlike Greenpeace or TWS, FoE argues that human social equity is a critical plank in the creation of an ecological society. FoE's sustainable societies Australia campaign, launched in 1997, was based on the concept of 'environmental space'. The key idea associated with this concept is that there is a limited set of non-renewable resources, which the Earth's people must share. Although FoE has always distanced itself from governments, it does take part in the peak council meetings (renamed the National Environmental Consultative Forum (NECF) under the coalition in 1996 before it was effectively cancelled) with the federal government, where representatives from the ACF, TWS, Greenpeace, WWF, HSI, state-based conservation councils and other major environmental organisations meet with the Environment Minister to examine policy.

During the new government's 'honeymoon' period in May 1996, it appeared to FoE that the Environment Minister, Senator Robert Hill, was trying to 'do the right thing' and that there was a 'genuine desire to get it right' (Walker 1996). To the NECF in November 1996, after eight months during which the government destroyed much of the environmental goodwill it had fostered prior to the election, the minister appeared to be confident that environmental issues were not a high government priority. Geoff Law (1996), who worked for Senator Bob Brown, described Hill's approach: 'His technique ... is to adopt the body language of someone bored and weary and to lower his voice to a lulling mumble as though the subject is ... so lacking in interest that he doubts anyone is terribly bothered about it'). The FoE representatives at the NECF also noticed the enthusiasm of May 1996 had abated:

> ... there was a distinct lack of commitment on behalf of the minister. The week was disorganised, papers late, we generally did not know what was going on. The minister appeared vague and non-committal during the meetings, constantly referring to his advisers and avoiding answering questions. He knows exactly what he is doing. (Moxham, Baker et al. 1996)

FoE was the only environmental organisation vociferously to oppose not only the linking of the partial sale of Telstra to the Natural Heritage Trust, but also the sale itself, and following its release of the December 1996 edition of *Chain Reaction* (no. 76) subtitled 'Government holds environment to ransom', FoE received no government funding in 1997. FoE had received approximately $19,000 in 1996 and the striking red cover of the December *Chain Reaction* became known as the $20,000 cover (Walker 1997). At the NECF meeting in 1997, FoE was threatened with being thrown out of what now was the government's green round-table discussions due to what the government construed as their overly critical stance.

At the national AGM in 1998, tensions began to emerge over the issue of the organisation's survival in an era of no government funding, with pressures also mounting from other national FoE branches with different operating styles. FoE remains, however, one of the most radical formal environmental organisations in Australia. Their loose confederation of branches more closely resembles a social-movement model of activity. Its promotion of regional self-determination and a network of active, if less numerous, members across the globe has enabled it to survive the recent downturn in membership that has struck other environmental organisations operating in Australia. One example is FoE Nouveau based in Adelaide. Brow (1997, p 53) writes:

> Friends of the Earth Nouveau in Adelaide have had a different experience of the 1990s. They have witnessed a gradual increase from 12 to 36 members. They have made the conscious decision not to put effort into a large scale member acquisition program. Instead, they acquire members based on word of mouth. FoE Nouveau has a small but active membership and attributes their gradual rise in membership to working on a local level, with a philosophy of direct action, but having the resources of an international organisation behind them.

Friends of the Earth is not a wealthy organisation, but due to its active and highly skilled membership it remains resource rich. Despite its recent reduction in federal funding, one suspects that this model of political activity will be more resilient in the long term.

The mini-boom in global environmental awareness and concern that the wealthier nations of the North experienced in the late 1980s reached its apex in the form of the Rio Earth Summit in 1992. If one lesson was to be learned from Rio, it was that green problems were international in character. International green organisations playing politics in Australia and overseas received huge levels of support.

Many of these organisations made financial commitments, manifested in a host of environmental campaigns and the employment of many additional personnel. Such commitments could not be maintained during the mid-1990s, when the rapid expansion in group membership ceased and then declined. Strongly centralised organisations such as Greenpeace went through particular hardships as funding sources dried up while outflows continued at the same high rates. This situation resulted in a change of emphasis for some organisations and sometimes produced infighting and turmoil as new ways of dealing with environmental issues had to be adopted to maintain the status and activity achieved in previous years.

Only one extensive, international environment organisation was left relatively unscathed: FoE (until the election of the coalition government in 1996). Due to its grassroots, low budget focus and rich base of non-monetary resources, FoE was able to maintain relatively stable structures and strategies while other groups struggled.

Although powerful forces to be reckoned with, international NGOs have not usually had the same access to governments as mainstream, national and state-based NGOs. This has changed somewhat in recent years, with Senator Hill prepared to deal with any organisation willing to do his bidding. International organisations such as WWF and HSI became favoured organisations.

Over the past 30 years, however, domestic organisations have tended to dominate domestic politics on environmental issues. This situation has been further intensified in Australia as environmental politics is still largely constructed within local, state and national frameworks, with little real emphasis placed on transnational characteristics of environmental issues. This insularity in environmental policy-making at government level is reflected in the performance of most Australian green NGOs. While many environment movements around the globe actively develop transnational campaign networks, their Australian counterparts often remain transfixed by issues within their own political boundaries.

NOTES

1   Notion examines the dominance of 'light green' values in the Greenpeace hierarchy, suggesting that this occurred because Greenpeace was 'just successful and got big. And they started caring more about their corporate fortunes than about changing the world.' She discusses the arrival of Steve McAllister in 1989 and describes Greenpeace's 'pyramid-selling environmentalism' money raising activities and its commitment to 'environmental theatre'.

2   Smith examines the public image of Greenpeace, how it has become very 'suit and tie', but also how its credibility has been called into question. His article includes an interview with Paul Gilding, former executive director of Greenpeace Australia, and discusses the business-like nature of the Greenpeace operation, including its money-raising activities: 'Complaints about high pressure selling tactics used by these door-to-door salesmen have dogged Greenpeace' and 'Everyone was there to make a few dollars, even the ones who really believed in Greenpeace.' Smith indicates the difficulty environmental activists have when confronting big business and demonstrates Greenpeace's need to market itself successfully to raise large revenues. This strategy backfired somewhat as expenditure on fundraising in 1995/96 almost eclipsed revenue (see 1995/96 annual report). Greenpeace has now effectively stopped this expensive method of fundraising.

NATIONAL AND STATE-BASED ENVIRONMENTAL NON-GOVERNMENT ORGANISATIONS IN AUSTRALIA

T hree organisations are the focus of discussion in this chapter. The Australian Conservation Foundation (ACF) and The Wilderness Society (TWS) are the most readily recognisable Australian green NGOs, and have been involved in many environmental campaigns. Both organisations involve themselves in environmental issues across the country, and they have membership in all Australian states, although membership is concentrated in the south-east.

Neither is a national federation of environmental organisations. Although often portrayed by the media as *the* voice of Australian environmentalism, the ACF is not broadly representative of the wider movement. It is better understood as providing leadership for the movement at certain times. Australia is without a national federation of formal, green NGOs. This role is fulfilled at the state level. One of these federations is reviewed here: the Conservation Council of South Australia (CCSA). Many of the issues that confront CCSA are indicative of the experience of similar federations in other states. Interestingly, over the past two decades the ALP has struck deals almost exclusively with the ACF and TWS, while the conservative parties have concentrated on negotiating with the broader based conservation councils and environment centres.

AUSTRALIAN CONSERVATION FOUNDATION

For the past three decades the ACF has been the most widely recognised national environmental player when it comes to dealing with the federal government. The ACF was formed in the mid-1960s and had its first council meeting in 1966. It was initially a very conservative body, with the late Sir Garfield Barwick, Chief Justice of the High Court and former Liberal politician, as its first president. In late 1968 the Commonwealth agreed to fund the ACF for $60,000 over three years, and by 1972 the conservative McMahon government had increased the ACF's annual grant to $150,000 (Warhurst 1992).

The ACF is a federation of state branches and chapters, but is fundamentally a Victorian and NSW organisation. Its council comprises five representatives elected from each state and two from each territory. These voluntary councillors decide the campaign priorities for the whole organisation. In addition, there are many professional staff who implement campaign strategies. There is always a tension between the elected officials and the professionals, but the ACF has often been riddled with abnormal levels of infighting as each group has sought to gain dominance. One reason for this conflict is that professional activists from other environmental organisations have too often become voluntary councillors of the ACF, bringing a professional and administrative ethos into council decision-making that directly confronts the 'amateur', more democratic traditions of the volunteers.

After the failure to prevent the flooding of Lake Pedder in the early 1970s (see chapter 8), many in the environment movement criticised the ACF's role as a 'collaborator' in the final decision. When a new council was elected in October 1973, Geoff Mosley was appointed director and most of the conservative councillors and appointed officers resigned. The ACF then became more activist and radical than it had been, but it was, and still is, seen as one of the more mainstream, moderate faces of the environment movement.

The ACF has been closely involved in a vast range of environmental campaigns, including Fraser Island, the wet tropics, Antarctica, and the damming of the Gordon-below-Franklin, but the ACF's involvement in governmental politics was relatively low-key until the appointment of Phillip Toyne, formerly a legal advisor to central Australian Aboriginal groups, as director in 1986. Toyne became the self-appointed mouthpiece of the environment movement during its period of growth and was a regular sight on the evening news until his departure in 1992. His role was critical in negotiating trade-offs with the federal government during this period and, partly as a consequence, in 1994 he was employed by the then Environment Minister, Senator Faulkner (Woodford 1995).

From 1992 until early 1996 Tricia Caswell served as the ACF's director. She oversaw a complete restructuring of the day-to-day running of the ACF that caused considerable dissent among the councillors and some members. When she took over from Toyne, the ACF's budget was in deficit and the organisation had to dip into a previously sacrosanct

$1.6 million endowment fund to keep it afloat. Former president Peter Garrett (who re-emerged as president in 1998) had warned Caswell that she faced a 'Herculean task to reform the organisation' when she commenced her tenure (Boreham and Milburn 1995).

Caswell had been a Victorian Trades Hall industrial officer and school teacher. It was believed she had the connections with the ALP needed to foster the close relationship Toyne had forged with the national government of the day. Caswell had little experience in environmental campaigning and was considered rather naive as an environmental activist. Caswell employed a professional financial manager and other professional staff, she relieved the council of its administrative duties and replaced volunteer campaigners, she centralised the organisation by closing branch offices, she shut down the disastrously unsuccessful 'merchandising arm', and she placed firm limits on the number of campaigns the ACF could involve itself in. Many of these 'executive decisions' did nothing to endear her to many in the council who were democratically elected from the various states and territories (Bita 1995).

In mid-1994 Caswell received a performance appraisal from a subcommittee of councillors saying they were very happy with her performance. In November that year, in a meeting from which ACF members and staff were barred, only four of the 37 national councillors supported her bid for contract renewal. The manner in which the meeting was held contradicted the open and democratic traditions of the organisation. Former director Geoff Mosley rejected pleas to legitimise the meeting by attending and declared it to be 'unjust' and amounting to a 'kangaroo court' (Boreham and Milburn 1995). Caswell remained in office until December 1995.

Following her departure, and in a determined change of direction, the ACF appointed experienced grassroots environmental campaigner Jim Downey (Casswell's partner at the time) as director. Downey was considered to have a more bottom-up style. He made the point that he considered Caswell's more managerial approach necessary at the time because of the funding shortages and reorganisation required during the recession period of 1991–94 (Bita 1995).

Prior to the Downey era the ACF had long held the view that government was the key agent of change (Simpson 1996a). Two factors have caused a change in emphasis. First, the election of a federal coalition government put pressure on the ACF to rely more on its state branches for campaigning and organisation. The coalition is much more state-oriented than the ALP, and the ACF's closeness to the ALP during its period in government seemed to have the predictable result of leaving the ACF 'on the outer' in dealings with the coalition (Walker 1996). Second, the globalisation of 'the environment' has, ironically, placed more emphasis on the local environment and less emphasis on strictly 'national' issues. The proliferation of local environment groups has given a local and sometimes more immediate voice to environmental issues.

Both these factors encouraged branch autonomy and the bottom-up approach, emphasised by Downey, with direct negotiations with business and communities. Unfortunately, most ACF branches had been gutted under Caswell due to her regime's centralist tendencies, its administrative and economic rationalist approach, and its close relationship with the ALP. The organisation was now floundering in the new political milieu. It was no surprise, then, that following the election of the Howard government it would face hardship, with loss of access to government and a reduction in funding.

The ACF has also responded in recent years to criticism of its focus on 'wilderness' issues. As we have seen, 'wilderness' campaigns are fraught with difficulties. The ACF had also been criticised for at least the impression that it excluded indigenous Australians from environmental politics and displaying a lack of concern for the rights of indigenous people in both 'wilderness' and urban areas. Both Noel Pearson, serving as director of the Cape York Land Council, and academic Marcia Langton connect the common use of the word 'wilderness' with the politics of terra nullius (Salleh 1996).

In an attempt to rectify this perception, one of the ACF's eight draft campaign priorities from the 1996/97 Annual Plan was 'to ensure that environmental protection for the Kimberley region is achieved in accord with indigenous aspirations through ACF negotiating and developing a regional agreement with Aboriginal organisations with conservation outcomes for the region' (ACF 1996, p 6). Ros Sultan of the ACF organised the 1995 Ecopolitics IX conference on Indigenous Management of Environmental Resources in Darwin. Although the conference itself was a departure from the accessible and democratic nature of other ecopolitics conferences, with papers presented by invitation only, it did represent a genuine effort on the part of the environment movement to integrate indigenous issues into the mainstream of the movement.

Following the Wik judgment in December 1996, the ACF became one of the most vocal environmental organisations in defence of native title on pastoral leases. In early 1997, with independent MP Pauline Hanson launching her One Nation party—and National Party backbenchers threatening to resign from the coalition government unless native title was extinguished by pastoral leases—John Howard was briefly in the unusual position of looking quite 'progressive' on the national political scene. In this climate, however, the views of environmental organisations such as the ACF went almost unheard in the mainstream media.

With the government so determined to extinguish native title in certain areas, the ACF and The Wilderness Society teamed up with the traditional peoples of Cape York in 1996 to form the Cape York Indigenous Environment Foundation, bypassing governments in their attempt to find common ground in green/black issues (Salleh 1996). This action is symptomatic of the environment movement's exasperation with governments and the desire to achieve results in the absence of influence with government.

In October 1997 Jim Downey stepped down as director of the ACF just half-way through his four-year term. Downey stated that he was severely disillusioned and exhausted with environmental activism. He quoted past 'failures' such as the anti-uranium, forests and Hindmarsh/Kumarangk campaigns as reasons for his departure. He had also been offered a spot on the Australian Democrats' Senate ticket.

Don Henry was appointed director in 1998. Henry had been director of WPSQ and had served on the seven-member executive of the ACF at the time of the 1987 federal election. In the years leading up to his appointment he had worked in the USA for the WWF. Henry is an extremely good 'people person' and when he took up the position there were attempts to unify movement concerns in a series of meetings at Mittagong (see chapter 1). Unfortunately, any trust-building between movement professionals was dashed by the EPBC debacle. Henry was operating in a period when the ACF (clearly identified with the ALP) was battling to remain a viable player.

The ACF now looks more positive and more resilient. In the final days of Caswell's reign the ACF set up Taskforce 2025. This taskforce was initiated to 'review the *modus operandi* and advise on goals for the next 30 years for the Foundation's work' (Brow 1997, p 55). The impetus of the review was the realisation that ACF membership was declining and ageing, and this has had quite severe financial implications for the organisation. The taskforce produced several options for the future, with most support coming from its members for Option Three. This was described as 'the Standard/Grass Roots Approach plus attacking the Problem at its Core' (Brow 1997, p 56). This approach attempts to devise a new economic and social system that gives the highest priority to the achievement of long-term sustainability. This is a far more radical direction for the ACF, reminiscent of broader social-movement goals. No longer does it see itself exclusively as a lobby group. It now seeks to extend its links both within the environment movement and in the broader society. To quote from Taskforce 2025 (Brow 1997, p 57):

> ... a social movement for fundamental social change will most likely involve the building of many alliances but they cannot be exclusionary.

Obvious targets for early alliances or sympathetic relationships beyond the environment movement are—the labour movement, indigenous peoples, human rights groups, green business, local government and local government associations and regional groups and agencies.

The ACF's strong recent support of indigenous groups over the battles for native title and Jabiluka is evidence of this new broad coalition-building aimed at more fundamental ecological and social change. The ACF continues to rely heavily on government funding and membership subscriptions. It has recently attempted to boost revenue through merchandising, bequests and payroll deduction schemes. Its administration has remained centred in Melbourne.

THE WILDERNESS SOCIETY

The Wilderness Society (TWS) grew out of the Tasmanian Wilderness Society, which was formed by Bob Brown in the mid-1970s during the campaign to save south-west Tasmanian wilderness areas. After becoming a national organisation in the early 1980s, TWS played a critical role in Australia's major wilderness disputes throughout the 1980s and early 1990s, being the instigator of Australia's biggest and most successful non-violent protest actions at the site of the Gordon-below-Franklin dam.

TWS DURING THE 1980s

The Full Bench of the High Court of Australia handed down a decision in 1983 that stopped the development of the Gordon-below-Franklin dam in south-west Tasmania (see chapter 7). Apart from saving this wilderness area, the decision had far-reaching effects on the Tasmanian Wilderness Society, which had been largely responsible for the success of the campaign. The goal that bound it together had been realised.

After the successful completion of the Franklin campaign, certain factions within TWS decided that the organisation had the resources and experience to transfer its efforts to other arenas of eco-activity. The 'Tasmanian Wilderness Society' became 'The Wilderness Society', and members of the society defined three major new goals: the Daintree rainforests, the national wilderness legislation and education program, and the western Tasmania national park campaign. There were concerted attempts to nationalise the society and, according to Holloway (1986, p 39), these efforts led to great debate over structure:

> With the decline in membership and the desire to retain the mobilising power of the organisation, TWS became increasingly formalised ... However all this had to be balanced against the desire to maintain ... the spontaneity, excitement and creativity associated with charisma and the egalitarian ideology associated with direct democracy.

At first it seemed that the contest may have been won by those advocating continued reliance on those informal, personalised relationships that had been largely responsible for the Franklin victory. It was a faction within TWS advocating non-violent action that supported the idea of Daintree becoming the political focus for the organisation. The NVA faction saw Daintree as a perfect opportunity to re-exercise their non-violent tactics and ideology. Those advocating NVA saw it as a goal in itself, quite separate from conservation of the wet tropical forests, though these two goals were seen as non-conflicting. Holloway (1990, p 6) writes that 'NVA is based on a wider program for reform (Movement for a New Society—USA) and includes consensus decision making as central to its anti-hierarchical critique of society'.

The situation changed, however, after the unsuccessful second Daintree blockade. Bureaucratic organisation started to reaffirm its primacy. This was aided with the resignation of TWS's charismatic leader, Bob Brown. There was an increased specialisation of tasks, wages were introduced, and new goals were modified by instrumentally oriented activities (Holloway 1986, p 38). Holloway's interesting dissertation concludes without deciding which form of political structure or ideology won out. He writes (1986, p 43) of various factions 'segmenting' to form interdependent parts still operating within the organisation, and poses the unresolved question: 'is segmentation merely a transition stage before relatively complete bureaucratisation or is it a genuinely new form of social organisation?'

The Weberian 'laws' of increased bureaucratisation and formalisation have no doubt exerted their influence on TWS. During the 1980s, TWS continued with the establishment of offices in most capital cities. A definite sense of competition evolved during this period between TWS and the ACF to be considered *the* national environmental organisation. In fact, several TWS members were among the key instigators in the move to sack long-serving ACF director Geoff Mosley. Mosely's removal as director occurred in April 1986, in the first closed meeting of the ACF. There is no doubt that during the 1980s growth, as a systems goal, seemed to be of prime importance for the society (see Table 6.1 for a breakdown of goal types). Other goals seemed to become much more conservative. For example, in the 1980s the once 'radical' TWS advocated the 'modification' of woodchipping, whereas the ACF demanded the phasing out of woodchipping altogether.

Once a Tasmanian organisation, TWS now involved itself in the intricacies of politics in other states. In Queensland, TWS maintained that its major goal was the conservation of the wet tropical forests, although concerted efforts from 1985 to mid-1986 seemed almost non-existent. Instead, in operational terms, it devoted time to all conservation issues that made their way onto the agenda. For example, it involved itself in the 1986 Queensland state election, in which the Queensland Conservation Council campaigned on the major theme of 'national parks'. Thus TWS dealt with the most reformist conservation organisations, and although it was still perceived as 'radical' by some groups, much of this was a legacy of its past.

This ideological dilemma still exerts its influence on TWS. What is unusual about The Wilderness Society (and to its credit) is that it has responded to and recorded the dilemma it constantly faces. Other environmental and social-movement organisations seem to undergo similar experiences but are either relatively ignorant of its existence, or simply perceive that they do not have the time or the inclination for serious soul-searching.

TABLE 6.1
GOAL CHARACTERISTICS OF THE WILDERNESS SOCIETY

1   Societal goals: Ambiguity within the organisation as to what these are. The NVA faction is concerned with revolutionary change towards a new, peaceful society. Bureaucrats are concerned with the maintenance of the current regime, and attempt to achieve change through legislative procedures. Although certain conservation goals—such as the Franklin and the wet tropical rainforests—have topped the agenda, TWS enters into many environmental issues.
2   Output goals: TWS has a broad range of output goals, indicating the diversity and ambiguity within the organisation. These goals manifest themselves in techniques used for mass mobilisation and lobbying elites.
3   Product goals: Because of the paradox that dictates so much of its operation, it is difficult to pinpoint TWS's product goals. Some factions wish to portray the organisation as the 'rip and tear', radical, accountable-to-no-one environment group. Since the success of the Franklin campaign, however, with the increased bureaucratisation of its affairs, its image has been more often respectable and politically expedient. In 1999 it aligned itself with those North American organisations that are involved with the 'Earth Charter' (see chapter 12), and now sells the business-friendly ideology of 'sustainable development'. This inability to label TWS's product goals with a clear political image may be of significant use to it in jockeying for power.
4   System goals: These are ambiguous due to the fact that many members are committed philosophically to egalitarian, open interaction, while others concede the necessity of increased formalisation to advance levels of efficiency and political clout.
5   Derived goals: There are many. It was shown that after the fulfilment of its original goal—the protection of the Gordon-below-Franklin catchment—TWS was eager to enter new areas of eco-activity.

Source: Following the criteria set out by G. Perrow (1970).

TWS FROM THE LATE 1980s TO THE LATE 1990s

Although its origins were in grassroots activities, TWS became intimately connected with the federal ALP government in the late 1980s, particularly around the time of the 1987 election. The election of Jonathon West as director of TWS prior to the 1987 election resulted in increased executive control and less input from the branches.

Partly in response to the member and volunteer backlash of this period (see chapter 10), TWS has since made a conscious effort to distance itself from government. Its change in attitude was demonstrated by its decision to boycott the Ecologically Sustainable Development (ESD) process of the early 1990s, which the Hawke government had set up, ostensibly to make development more environmental-friendly, but effectively to diffuse the conflictive nature of environmental issues that had arisen in the 1980s. The ACF and the World Wide Fund for Nature (WWF) participated in the ESD process, while Greenpeace Australia was involved in the initial stages but later withdrew (see chapter 9).

The relationship between TWS and the government had cooled following the appointment of Ros Kelly to replace the influential Graham Richardson as Environment Minister in 1990 and Keating's successful leadership bid in 1991. Following the ALP's successful election campaign in 1993 without obvious environmental support, the Keating government felt it could essentially afford to alienate TWS, which Keating tried to paint as '"Goebbels-like" extremists' (Boreham and Milburn 1995), without apparent electoral repercussions. During the row over woodchip export licences in 1994–95, veteran TWS forests campaigner Alec Marr withdrew from negotiations after being frozen out by the federal government. This action followed accusations by then Environment Minister, Senator John Faulkner, that Marr had reduced government officials to tears over forestry policy negotiations (Bita 1995).

TWS then started using Virginia Young, the somewhat quieter Queensland coordinator, for its dealings with the government. As TWS media officer Felicity Wade explained: 'We're recognising the Government's out to get Alec and [former campaign director] Kevin [Parker], as the boys who do the big argumentative "rah rah rah"' (Bita 1995). Since the early 1990s TWS has undergone a significant shift in overt policy direction, with the June 1993 national meeting formally acknowledging that TWS's direct-action campaigning ability depends, to a significant extent, on TWS members and supporters across the country (Lambert 1994). This is in direct contrast to the top-down approach that emerged during the 1987 election when supporters were instructed to campaign, not for wilderness, but for the ALP.

Having tried for years, like the ACF, to do too much at once, TWS set up an Interim Campaign Committee to define a new campaign structure driven by branches. The committee focused on six aspects of wilderness campaigning work: forests, Cape York Peninsula, wilderness education, outback (arid and semi-arid) wilderness, Commonwealth wilderness protection, and international wilderness. Working groups were then set up to deal with each issue (Lambert 1994).

Although there is not a huge emphasis on social or 'brown' (urban) issues in this campaign list, ex-director Kevin Parker insists that TWS has a very socially oriented basis due to its roots in the peace movement (Simpson 1996b). Its commitment to the peace movement and associated philosophies is more closely related to that espoused by FoE than by the ACF or Greenpeace. For instance, the importance of consensus decision-making is emphasised in 'The Management of TWS Meetings' in section 5.4 of the TWS Policy and Procedure Manual (TWS 1994), as is the requirement for a branch to conduct a workshop in the philosophy and practice of non-violence every year. It is also obvious, by name and by campaign list, that it still concentrates on wilderness and ecocentric issues. This narrow focus is demonstrated by TWS's inability to entertain a policy on the linking of the sale of Telstra to the Natural Heritage Trust, which the coalition took to the 1996 election, due to the fact that TWS's 'charter does not extend to engaging in debate regarding public asset sales' (Parker 1996).

Leading up to this election TWS endorsed neither the ALP nor the coalition, but it was quite clear in Parker's 'From the Campaign Coordinator' columns in the TWS newsletter, *Wilderness News* (Parker 1996), that it considered the ALP's environment package inferior to the coalition's. Furthermore, in January 1996, TWS issued a press release in which it:

> welcomed the environment statement from the Coalition ... The prospect of working constructively with a Coalition government is now very possible ... Overall it is a good day for Australia's environment movement because we now have a broad bipartisan approach on many issues and the greater commitment of the Coalition has put the [ALP] government in the shade. (TWS 1996)

TWS was clearly signalling that the days of automatic support for the ALP were over. As early as 12 months before the 1996 election, TWS had approached the Liberal Party about forming a relationship in case the coalition won the election (Shaw 1996). As a result of this strategy, and the ACF's closeness to the ALP, the situation was reversed, with TWS gaining more access to the coalition government and the ACF being left on the outer (Walker 1996).

A problem arose when the coalition in government proved itself to be less than committed to the environmental issues it had supported prior to the election. While TWS came out publicly in opposition to many of the coalition's policies in government, Marr stressed that dis-illusionment with the coalition would not automatically turn into sup-port for the ALP. The ALP would have to come up with good environmental initiatives if it wanted to regain the position of favoured party of the environment movement. Marr suggested that environ-mental policy, in general terms, could have been even worse under a new ALP government (Shaw 1996).

With neither major political grouping offering acceptable options, TWS has once again had to return to its grassroots approach and local actions, with the government often being bypassed in favour of direct negotiations with business and other interest groups. TWS closed its Canberra office after the 1996 election campaign. This strategic turn was a move for the better. TWS has achieved much over the years in its mass mobilisation campaigns. Its many forays into electoral politics, however, have been nothing short of disastrous.

Of the four organisations referred to in this chapter, TWS has expe-rienced the most serious decline in membership. Although many in the organisation blame a lack of continued support by the Australian pop-ulation, others within TWS accept that mismanagement has also con-tributed to this decline. During the 1990s TWS has 'experienced high staff turnovers and internal inefficiencies which have been compound-ed by a lack of internal review' (Brow 1997).

CONSERVATION COUNCILS

There have been attempts in all Australian states to build federations of different environmental organisations. Most of these federations were brought together in the late 1960s and early 1970s. Originally, they provided services to organisations that tended to pursue more tradi-tional nature conservation goals, though the net has been cast more broadly in recent times. Most of these councils are 'broad churches', with organisations from both the left and right of the political spec-trum, and many other places between and elsewhere! These councils are as close as possible to being institutionalised versions of social movements. Herein lies their true fascination. Although councils are not as diverse as a full social movement, given that they include only formalised groupings, their diversity can nonetheless be quite startling when compared to many other organisational forms. A social move-ment, of course, could never be fully institutionalised, as it would then cease to be a social movement.

Table 6.2 lists the organisations that gather under the umbrella of the Conservation Council of South Australia. In the CCSA's 30-year history, strange and wonderful bedfellows have operated under its rubric. For example, hunting organisations have existed alongside ani-mal welfare groups, right-wing population control organisations have jockeyed for position with left-wing groups fighting for social and environmental justice, and wilderness and nature conservation groups have functioned together with groups more interested in the human environmental dimension. Some of these tensions have resulted in conflict, but they have usually led to a creative and broad power base that provides these organisations with their major strategic weapon: the power

FIGURE 6.2
MEMBER ORGANISATIONS OF THE CONSERVATION COUNCIL
OF SOUTH AUSTRALIA

Action Group for the Protection of Coffin Bay Waterways
Adelaide Bushwalkers
Adelaide Hills Environment Association
Anthropological Society
Association of South-East Field Naturalistsí Societies
Australian Conservation Foundation
Australian Marine Conservation Society (Adelaide Branch)
Australians for an Ecologically Sustainable Population — SA Committee
Bicycle Institute of South Australia
Bird Care and Conservation Society
Civic Trust of South Australia
Coolabah Club
Echidna Care
Eco-Action Kangaroo Island
Economic Reform Australia (SA)
Field Geology Club of South Australia
Field Naturalists Society of SA
Friends of Goolwa & Kumarangk
Friends of Port Moorowie
Friends of the Tea Tree Gully Hills Face & Rural Living Zone Inc
Friends of Willunga Basin
Gawler District Environment and Heritage Association
Gould League of South Australia
Greenpeace
Highbury Environs Against Refuse Tips (HEART)
Institute for Earth Education

Marine Life Society of South Australia
Mt Barker District Environment Assoc.
Mt Lofty Ranges Conservation Assoc.
National Trust of South Australia
Natural History Society of SA
Nature Conservation Society of SA Inc
North East Hills Environment Conservation Association
Orienteering Association of SA
People for Public Transport
Permaculture Association of SA
Port Adelaide Residents Environment Protection Group
Royal Zoological Society of South Australia
Scientific Expedition Group
Society for Growing Australian Plants
Soil Association of South Australia Inc
South Australian Herpetology Group Inc
South Australian Ornithological Assoc.
Southern Districts Environment Group
Spencer Gulf Environment Alliance
St Agnes Bushwalking and Natural History Group
Toyota Landcruiser Club of Australia (SA)
Trees for Life
Urban Ecology Australia Inc
Walking SA
Wilderness Society (SA Branch)
Yookamurra Society Inc

A 'critical mass' bicycle ride through the streets of Brisbane. *Phil Bradley*

of diversity combined with solidarity. CCSA officially connects 60 green organisations (with approximately 60,000 individual members) and many other community groups and more informal networks and associations that use its central facilities.

The role of conservation councils is often so large and ill-defined that it can be quite unrealistic. CCSA (1998) lists its role as follows:

a) provision of a cooperative forum and resource base for its member organisations;
b) provision of information and educational services to the general public;
c) provision of a link between the community, government and industry;
d) facilitation of public participation in community environment projects.

This list can really be reduced to two main tasks: to provide its member organisations (as well as other green groups) with services that allow them to pursue their own narrower objectives, and to provide political leadership on certain key issues. Sometimes these two goals are in direct contraposition to each other. The organisation attempts to be democratic, being driven by the wishes of its members, but at the same time is expected to deliver centralised policies that can influence government decisions.

This is just one source of tension in these broad-based formal coalitions. Like most environmental NGOs, these councils have a voluntary board, most of whom are members and representatives elected from the member organisations. Ultimately, the responsibility of the elected board is to the member organisations they represent. The board employs professional environmentalists to pursue its policy directions and administration, while most campaigns are still fought by volunteers.

Imagine you are a professional employee of one of these councils. You are an ardent environmentalist operating in poor working conditions, are poorly paid and work longer hours than your contract specifies. Despite these conditions, you have many constituents whom you have to please to keep your job and to continue professionally fighting for the cause that is so important to you. These constituents include: the 60 affiliated organisations, their members and representatives;[5] the voluntary board of the conservation council; the board executive, which is a smaller collection of key councillors and the CEO; your CEO (who is the chief professional in the organisation); and a huge number of volunteers who make up the person-power of the organisation. Apart from these constituents, you also try to remain on good terms with bureaucrats and politicians, corporations, trade unions and other community organisations and coalitions. No other sector places such enormous political and workplace pressures on its key professionals.

Given their complex task, conservation councils usually remain non-party-political, and due to their very nature, it is essential that they do so. Individual environmental organisations such as the ACF and TWS are involved in the delivery of votes to particular parties. Federations, on the other hand, usually have a longer time frame when analysing their political role, so their political affiliations are more reminiscent of a bureaucracy or institution that has to operate under any elected regime. Interestingly, they are constantly pressured to become involved in the electoral process by political parties and by specific environmental organisations. The example of the 1998 federal election is of interest here.

THE 1998 FEDERAL ELECTION

Despite having been made aware of the pitfalls of being identified with a particular political party, the ACF took an active though often uncertain and ambiguous role during the lead-up to this federal election in supporting the ALP. The ACF took it upon itself to coordinate a broad, national network of environment groups, including the state-based peak organisations such the CCSA. The initial argument by the ACF and a self-appointed network of professionals from these organisations was that a joint strategy of major environment organisations would avoid duplication of effort and be more efficient. This is a fair claim, as elections place great demands on NGOs.

The early process was oriented around a US-style electioneering 'report card', designed to identify objectively the 'best' party for the environment. Once members of the public had weighed up the various categories, then they would vote appropriately, without the green electoral coalition having actively to endorse a particular party. At first glance, the ACF and others appeared to have learned from past mistakes, but as the process went on its flaws and biases became apparent.

First, the questions asked on the report card are more important than the answers. The questions could only lead to answers that would denigrate the Howard government's performance and solicit support for Labor and the minor parties and independents. As I will argue in chapters 11 and 12, the performance of the Howard government on environment issues has been appalling. What is of interest here, however, is the ACF's slightly more subtle way of playing electoral politics.

Just before the election, this subtlety disappeared. One week before the election, the ACF released a document entitled 'Nature Heritage Trust: What happened to that $1 billion for the environment?'. In this report, the ACF attacked the government and identified major environmental cost shifting. Environment Minister Robert Hill (1998a) wrote to peak groups complaining of this ACF broadside:

> Unlike many other environmental groups around Australia, the ACF has shown little interest in the management of the Natural Heritage Trust Programs over the past two years. To come out one week before the Federal election with this political stunt can only further damage the credibility of an organisation that was once a leading advocate for conservation in Australia.

The National Heritage Trust (NHT) was a political hot potato for the coalition early in 1998. The government was placed under intense scrutiny regarding its distribution of funds under the trust. 'Green-barrelling' was a term that emerged from the rhetoric of the ALP opposition: a major aspect of the critique being that the Commonwealth's funds were being directed disproportionately to coalition electorates. Most of the funds did go to Liberal or National electorates. This may have had less to do with a conscious and strategic manipulation of the public purse by the Commonwealth than with the rather narrow definition of the symbol 'environment' that has evolved within coalition policy.

The *green* agenda during this third period of modern environmentalism has been recoloured as a *brown* one (see chapters 11 and 12). The important issues of land clearing and degradation have assumed primacy, while the myriad other environmental concerns, such as those relating to urban centres or forests, have been placed firmly on the back-burner. Consequently, most of the 'environmental' monies derived from the partial sale of Telstra were directed towards rural, primary industries. Obviously, with Labor's electoral heartland being urban-based, little of this funding wound up in Labor electorates. Two other problems with the NHT process should be mentioned here. First, in the Commonwealth's pursuit of short-term results, on-ground work has become the mantra of the trust. The urge to act has resulted in many trees being planted. However, without sufficient attention to, and funding for, research, facilitation and strategic planning these results were piecemeal and of limited benefit. Second, there have been serious problems with the state governments' involvement in this Commonwealth process. The key concern here is that states were using NHT funds to pursue projects that should be funded from normal sources. In addition, there have been concerns with the states' undue influence in the selection of the membership of the regional and state assessment panels. In this manner, supporters of biodiversity have been overshadowed by those more concerned with 'sustainable agriculture'.

Through the ACF's whole-hearted adoption of an issue that had been used by the ALP to play political ping-pong with the government, it tarred itself with the Labor brush one week before the election, as it has done on numerous occasions over the past 15 years. Some would say that this position is the ACF's prerogative. Unfortunately, by portraying itself as representative of the wider environment movement, there would be significant repercussions for other organisations loosely associated with its clumsy attempts at delivering the ALP the 'green vote'.

CCSA was extremely concerned that this ploy might involve it in an electoral strategy that would connect it to the ALP. It formed its own electoral subcommittee, which argued that as conservation councils are federations (including the ACF), they have a broader constituency than any single organisation. The CCSA decided to withdraw from the electoral network led by the ACF, and reiterated its position that its task was not to deliver votes for political parties. This position did not exclude CCSA from working with or against any political party on a particular issue between elections.

Several consequences resulted from this decision. First, there was some alienation from the other peak groups that remained in the process. Second, in not endorsing the ALP, the CCSA was immediately labelled as supporting the coalition. Party politics at election times are so powerful that the media, among other players and analysts, cannot present politics in any way other than 'for' or 'against' major parties. Advocates of longer term social-movement styles of politics find it difficult to retain their focus and power. One positive repercussion of CCSA's non-party-political stance was the maintenance of its federal funding levels for 1999, while the ACF, not surprisingly, again suffered significant cuts. With the green electoral network forcing the CCSA publicly to restate its non-party strategy, there may be further repercussions in the future when the ALP returns to government.

Unfortunately, party politicians do not sufficiently understand or respect the nuances of movement politics. As a result of the ACF electoral network's efforts, most green organisations were seen as having supported the ALP during this election, whatever their individual stances, and this was reflected in another round of cuts to community-based environmental organisations in the 1999 federal Budget.

This electoral strategy of mainstream green organisations was not in the same league as the highly orchestrated efforts displayed during the federal elections of 1983, 1984, 1987 and 1990. In 1998 it was not carefully planned; rather, it was shabby and ill-considered. Many strategists in the network were not committed and felt ambivalent about the process. In a rather embarrassing meeting just weeks before the election, the ALP informed the green electoral network that it did not want the green vote anyway, as the election was 'going to be won in the bush'.

Over the past two decades the ALP has worked closely with the ACF, while the conservative parties have sought the aid of conservation councils in service delivery (and now some hand-picked international organisations). While the ACF worked tirelessly on the ESD process, the councils were largely uninvolved. Similarly, the councils are the National Heritage Trust's major infrastructure provider at the community level, whereas the ACF has been largely absent from the process.

The larger constituency of the conservation councils provides solidarity, as well as access for its member bodies to key policy-makers. On the negative side, conservation councils such as the CCSA are often criticised for not taking a strong stance on particular issues. Having to take into account so many opinions can induce conservation councils to produce conservative, uncontroversial goals. It is no great surprise that these councils and centres continue to attract federal and state government funding due to their often innocuous stance on many issues. What is genuinely surprising, however, is that some truly remarkable people have survived and flourished under these conditions, managing to produce some excellent environmental policy outcomes in the medium to long term.

The last three decades have seen momentous changes in the atti-

tudes, funding, structures and perceptions of environmental organisations in Australia. The formal organisations reviewed in the last two
chapters have faced hostile governments, industries and other adversaries. What is remarkable about these organisations is their apparent
resilience. They have survived, and sometimes thrived, in very different
policy-making eras. Sometimes they have been sought out by governments, and on other occasions they have seen their funding reduced.
Usually they have provided viable, sometimes confused but alternative
pathways for citizens to contribute to our political lives outside traditional partisan politics. Part two offers a loose chronology in an
attempt to link these many participants in the environment movement
over a generation.

NOTES

1   The story of the 'merchandising' arm of the ACF has never been fully disclosed.
    Carried away with the euphoria of the later 1980s and very early 1990s, the ACF
    began to take ridiculous risks with what were public assets. One disastrous attempt
    to develop and market a low-energy light bulb led to unprecedented debts. In the
    early 1990s the debt of the organisation's corporate arm was in the millions. Much
    of the recent literature about the ACF talks about the decline of the economy and
    its impact upon membership numbers. But there is rarely any acceptance that
    some fiscal restrictions have occurred because of mismanagement and the misuse
    of its members' assets.
2   Professional environmental activists really do constitute a subculture unto them-
    selves. The incestuous nature of these networks is legendary and, on occasions,
    really does raise the question of nepotism. As we saw in chapter 2, their networks,
    private friendships and alliances are defining factors in explaining decision-making.
3   Salleh writes: 'The concept of wilderness has become a site of contestation and
    struggle between eco- and Aboriginal activists. But higher synthesis is possible'.
4   Although Howard appeared to be taking the middle ground, it was obvious from
    such comments when referring to Hanson as 'she has no cure to the disease' (ABC
    1997), that he believed there was some 'disease' in our society that Hanson had
    identified.
5   The memberships of each of the conservation councils and environment centres
    alone—existing in each state—far outweighs the membership of all the political
    parties in Australia put together.

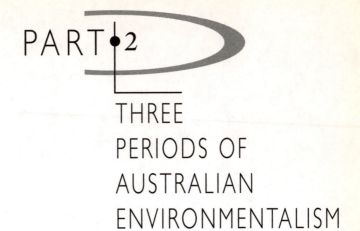

PART 2

THREE
PERIODS OF
AUSTRALIAN
ENVIRONMENTALISM

7

THE FIRST PERIOD:
UNRESTRAINED USE

In this chapter and the next we look at the first period of modern Australian environmentalism, that is, at developments between the late 1960s and approximately the mid-1980s. This period was dominated by conservative federal governments, sandwiching the Whitlam years of 1972–75. The blurred edges of the three stages are deliberately contrived. Although this period ends with Hawke's ascension at the federal level in 1983, his consensus-style policy-making took some time to fully grip mainstream movement practices.

We examine first the ideology of unrestrained use as a dominant discourse. I am not arguing that the ideology of unrestrained use (or any other ideology described over the next chapters) constitutes a coherent and clearly constructed manifesto that delineates the views of the state from those of the movement. What is, and what is not the state is as open to conjecture as what constitutes the ideological underpinnings of the state. These characteristics change over time, and are cast differently in separate theoretical understandings of power and the policy process. Naming distinct ideologies simply provides a working description of some of the constellation of ideas, myths, beliefs and themes that are part of what Thomas Kuhn refers to as 'the dominant paradigms' (Kuhn 1962). Some components of paradigms, of course, change incrementally. Those paradigms championed by institutions usually change more slowly than others.

We then examine the beginnings of the movement that formed largely in opposition to this ideology. This movement is characterised by playing two styles of 'outsider politics': one based on the unquestioning acceptance of pluralist models of power, and the other on aggressive and structuralist opposition to prevailing practices. We then turn to pluralist and structuralist depictions of the state and policy-making processes.

THE IDEOLOGY OF THE STATE: UNRESTRAINED USE

In his book *Toward a Transpersonal Ecology*, Warwick Fox lists six key characteristics of what he terms 'unrestrained exploitation and expansionism' (see Table 7.1). This list of attributes is an excellent shorthand version of what I have termed 'unrestrained use'. The unrestrained use view of the world, with its clear division between humanity and the 'rest of nature', has its origins in ancient Greece: we often refer to it as the Western view of nature. It is anthropocentric, as non-human nature is only seen to have value when it has use for humans. When coupled with the technological advances since the scientific and industrial revolutions, this marked dualism has led to extreme planetary degradation. The ideology is short-term, as it does not take into consideration the impact of such a negative relationship on future generations, and it is blinded by technological optimism.

Points 2 and 3, which Fox refers to as the myths of progress and superabundance, demand more attention here. Myths often take the form of maxims or clichés. Each myth is very much part of 'common sense'. Boudin (1988, p 1) defines 'common sense' as what might be interpreted as a 'collection of myths':

> I will use the complex notion of common sense in a narrow way and with, I hope, a precise meaning. I will designate by this expression the fact that in our everyday life we use concurrently a number of principles 1) to which we grant a universal value 2) of which we are hardly conscious. Because they appear to us as natural, as universally valid, we tend never to put them to question: like the familiar objects which surround us, we perceive them in a *metaconscious* fashion.

The myths of progress and superabundance are integral parts of this 'common sense' relationship between humans and other nature.

THE MYTH OF PROGRESS: 'YOU CAN'T STOP IT'

> The History of the Human Species as a whole may be regarded as the unravelling of a hidden Plan of Nature for accomplishing a perfect state of Civil Constitution for Society ... as the sole State of Society in which the tendency of human nature can be all and fully developed. (Kant 1984)
> Poetic imagination has put the Golden Age in the cradle of the human race, amid the ignorance and brutishness of primitive times; it is rather the Iron Age which should be put there. The Golden Age of the human race is not behind us but before us; it lies in the perfection of the social order. Our ancestors never saw it; our children will one day arrive there; it is for us to clear the way. (Saint-Simon 1814)

TABLE 7.1
UNRESTRAINED USE CHARACTERISTICS

1   It emphasises the value to humans that can be acquired by physically transforming the non-human world (e.g. by farming, damming, mining, pulping, slaughtering). It emphasises, in other words, the physical transformation value of the non-human world.

2   It not only measures the physical transformation value of the non-human in terms of economic value, but also tends to equate the physical transformation 'resources' with economic growth. Economic growth is then equated, in turn, with 'progress'.

3   In order to legitimise the continuous expansion of resource exploitation (physical transformation) activity, this approach relies on the myth of superabundance, that is, on the idea that there is 'always more where that came from'. In view of this feature in particular, some writers have characterised this approach as frontier or cowboy ethics. Unrestrained exploitation and expansionism—frontier or cowboy ethics—is 'how the West was won'.

4   It is totally anthropocentric; the non-human world is considered to be valuable only insofar as it is of economic value to humans.

5   It is characterised by short-term thinking, that is, its anthropocentrism does not even extend to consideration of the interests of future generations of humans. This can be viewed as an expression of the radical anthropocentrism of this approach since one can afford to ignore the possibility of future problems if one has an unquestioning faith in the capacity of human ingenuity to meet these problems as they arise.

6   When forced to consider the longer term, deleterious effects of continued unrestrained exploitation and expansionism—or continued 'business as usual'—this approach falls back on technological optimism (i.e. on a faith that 'technological fixes' will always be able to deliver us from possible harm).

Source: Adapted from Fox (1990).

Towards the end of the seventeenth century there was much debate about a supposed 'golden age' in human history. The scholars of ancient times had previously dominated philosophical discussion. An integral part of this discussion had been a strong yearning for a past, noble age where society had reached ultimate fulfilment. As J.B. Bury writes in his *The Idea of Progress* (1960), philosophers such as Kant and Saint-Simon led the 'moderns' to a victory over the 'ancients'. For them, the 'golden age' existed in humanity's distant future; a future towards which humanity must strive. Krishnan Kumar (Krishnan 1978, p 14) writes: 'Mankind could now be seen as advancing, slowly perhaps but inevitably and indefinitely, in a desirable direction. In a sense, it was illogical to try to determine the happy end-point of this progression; but the attraction to do so proved irresistible.'

The scientific revolution had convinced certain intellectuals that new ideas and new tools had given humanity a chance to fulfil itself beyond past expectations. Modern humanity had reason, liberty and science. What were those 'primitive' societies in comparison with this great intellectual and social wealth? Humanity was now more 'complex' than ever before. The idea of progress has become extremely powerful. It is also mythical in every sense. It is seen as something beyond human control: progress is inevitable; one cannot stop it. To a believer in the myth, progress always existed; it just happens to be part of the 'natural forces'.

Progress, as a concept, became a dangerous myth in the context of the environmental crisis when, as Bury writes, it was linked with the Baconian idea that knowledge should be applied to human manipulation of the natural environment. Progress came to be seen as intrinsically tied to changing, taming, and controlling the non-human world. One aspect of this view was the idea that nature could be improved upon. Many environmentally degrading acts are now carried out under the mythical banner of 'progress'. The question remains: progress in what, to what?

> The future beckoned urgently, and the promise it held out could only adequately be gauged by the chaos that might result if the forces of progress were not all combined in the task of bringing the new society into being. Of those forces the most important were science, the men of science, and all those who could see in the achievements of the scientific method the highest fulfilment of the Enlightenment, and the key to the future direction and organisation of society. (Krishnan 1978, p 26)

THE MYTH OF SUPERABUNDANCE: 'THE UNIVERSE IS INFINITE'

> At that moment another dismal scream rent the air and Zaphod shuddered.
> 'What can do that to a guy?' he breathed.
> 'The Universe,' said Gargravarr simply, 'the whole infinite Universe. The infinite suns, the infinite distances between them and yourself an invisible dot on an invisible dot, infinitely small.'
> 'Hey, I'm Zaphod Beeblebrox, man, you know,' muttered Zaphod trying to flap the last remnants of his ego. (Adams 1980, p 75)

Fox (1990, p 152) refers to superabundance as articulating the view that there is 'always more where that came from'. There is always another frontier somewhere, for the earth is a limitless cornucopia, and the universe is infinite. In mathematics, infinity is a concept that explains increase without bounds. James Thompson—a mathematician—goes to some lengths to castigate the common-sense notions of infinity, as 'can be seen in the Oxford English Dictionary'. He interprets the *Oxford* definition as follows:

> ... something is infinite if it has no limit or end, either real or assignable, or is boundless or unlimited or endless, or is immeasurably great in extent or duration or in another respect. The term is used chiefly, it says, of God and his attributes, but also of space and time; here the use passes into the mathematical. Of this, the dictionary says that a quantity is infinite if it has no limit or is greater than any assignable quantity and that a series is infinite if it can be indefinitely continued without ever coming to an end. And these phases are themselves in need of explanation ... Now it is a quite intelligible suggestion that there are things which exist in infinite quantity, for example, that there is infinitely much gold in the universe. But the account suggested of what this means is absurd; a quantity cannot be greater than it is, and surely in saying that there was infinitely much gold we would be assigning a quantity to the gold. (Thompson 1967, pp 183–4)

If Thompson's point is valid, then the common-sense notion has no real mathematical or logical basis. As we have seen, this is a key characteristic of myth. The common-sense, mythical notion of an infinite universe has its roots in modern theology rather than in mathematics. The infinite, perfect God of the theists is the source of the idea in its modern usage. In this interpretation, God's infinity is formless and incomprehensible to finite beings such as humans. Despite the fact that we cannot imagine an infinite God and his or her universe, we 'know' it exists, for, 'if the universe is finite, then what exists beyond it?: there must be something!'.

How does this myth relate to the ecological crisis? The argument is that if the universe is limitless, then—given the appropriate technology—resources are also limitless: 'If we mess up here, then we can always move onto the next place.' Most modern views of capitalism remain firmly entrenched in the Lockean premises of property and abundance constructed in the late seventeenth century.

Probably one of the earliest references to limits to growth after the scientific revolution came from Hobbes (1914). Although Hobbes regarded the natural environment as limited, this was only due to the human incapacity to exploit resources. To him, nature, as an idea removed from human endeavour, was still limitless. It was not until Malthus's theories revolving around mathematical constraints on population growth that the finite nature of the environment was recognised (Flew 1970). In many non-Western cultures, these ideas of limits to growth have existed for millennia.

In Western post-industrial societies, however, the myth of an infinite universe with its infinite resources still reigns supreme. This idea may be further promoting the environmental crisis.

'What do you think will happen to them all?' he said after a while.

'In an infinite Universe anything can happen,' said Ford. 'Even survival. Strange but true.' (Adams 1980, p 197)

The extremism of the dominant ideology of unrestrained use, coupled with its resultant devastation, led to the emergence of the modern environment movement, which formed in dramatic opposition to what it perceived as an environmental crisis. Since the emergence of the movement, the state has fine-tuned its overarching ideological and management frameworks under new titles such as sustainable, multiple, wise and sequential use. These more recent conceptual frameworks are often no more than fine-tuning to the ideology of unrestrained use, which is so entwined with the Enlightenment project that it is inseparable from the very fabric of advanced industrial societies.

SOME BEGINNINGS, SOME PEOPLE, SOME EVENTS

No doubt, some well-meaning historian in a future time will pinpoint it with great accuracy and describe its 'inevitable' emergence as a force in Australian political life. For those of us, however, who have had too many meetings, too few co-workers, too many deadlines and too little money, there has been little time to think about where we came from or the 'inevitability' otherwise of our emergence. (Hutton 1987, p 2)

It is extremely difficult to talk of beginnings. Clearly identifiable causative moments do not exist. The historian living in the present projects into the future and looks back at her subject: time and interpretation are intermeshed. But certain things are self-evident about the early days of the movement in Australia. The first is that it did not suddenly jump onto the national political stage in the 1960s.

The environment movement, though assuming nowhere near its current power, has existed since the nineteenth century. Its emergence as a substantial political force grew slowly throughout the twentieth. Some writers identify different 'waves' of environmentalism, and weld these to some chronological structure (see Dowie 1995; Hutton and Connors 1999). Indeed, Hutton and Connors document this early Australian movement, devoting the first half of their book to the two waves of environmentalism that occurred before the emergence of the 'modern movement': the first, from 1860 to the Second World War; and the second, from 1945 until 1972. My own work concentrates on the period known as modern environmentalism or post-environmentalism (Young 1990), from the late 1960s or early 1970s until the present day.

Several seminal works from the USA gave impetus to this 'third wave'. For example, Rachel Carson's *Silent Spring* (1962) argued that indiscriminate use of pesticides was poisoning every part of the food chain. Futurists such as Alvin Toffler (in *Future Shock*, 1970) argued that mainstream, industrial society was pursuing a dangerous course, which would belittle and alienate human lives. Similarly challenging to existing unrestrained use orthodoxies was Paul Ehrlich's *The Population Bomb* (1972). It roughly coincided with the 1972 Club of Rome report, which reintroduced Malthusian notions of limits to growth to mainstream politics. Ehrlich and the Club of Rome argued that the Earth could not always provide sufficient resources for its people considering the current rates of population increase and the capacities of industrial production.

Barry Commoner, another American, came up with 'Four Laws of Ecology' in 1971: (1) 'everything is connected to everything else', (2) 'everything must go somewhere', (3) 'nature knows best', and (4) 'there is no such thing as a free lunch' (Commoner 1971). These neat maxims based on a new scientific interpretation of ecology assumed some prominence in the Australian movement The first two and the fourth laws were really part of one idea: interconnectedness, while the third law constituted a myth in every sense as it gave ultimate knowledge to a powerful force that is beyond human perception: the theists' God had been replaced by Nature with a capital 'N'.

Also important, however, were some key local events, and some rather remarkable people who helped shaped the Australian experience. It is perhaps unfair to single out people and campaigns, but I want to mention some because they proved inspirational in the emergence of the Australian movement. My selection is not intended to diminish the contribution of many others.

The advent of the campaign to save Tasmania's Lake Pedder and the evolution of the world's first green party—The United Tasmania Group—was a key shaping moment. People such as the late Dick Jones, who was central to the establishment of the Centre of Environmental Studies at the University of Tasmania, will always be remembered in this context (see chapter 8). One day the beauty of Lake Pedder may emerge again from its watery grave. Other Tasmanians have provided the movement with leadership. Bob Brown, for example, was the central figure of the Franklin Dam blockade and the Tasmanian Wilderness Society (now The Wilderness Society). He now leads the Australian Greens.

Others advance the idea of a movement conceived in Queensland (see Hundloe 1985; Wright 1977). Judith Wright has given us her poetry and her spirit. Her involvement in the 'coral battleground' of the Great Barrier Reef was instrumental in its ensuing protection. She was also a founding member of the Wildlife Preservation Society of Queensland (WPSQ), and was one of the first white Australians actively to campaign for the rights of indigenous people, an issue that many white 'wilderness' advocates have sorely neglected. Len Webb, also a member of WPSQ, was the son of a shearers' cook, became a world-renowned rainforest ecologist and was instrumental in many campaigns. His knowledge, humour and inspiration provided a bedrock on which subsequent campaigns to save the wet tropical forests were built. John Sinclair, who was central to the Fraser Island Defenders Organisation (FIDO), literally gave everything—all his money and his private life—in his head-on battle with the Bjelke-Petersen government to conserve the magnificent Fraser Island.

Many 'alternative lifestylers' moved from the big cities in the 1970s in search of more meaningful, fuller lives (Metcalf 1984). They personalised their political actions by incorporating major changes into the way they lived. Nan and Hugh Nicholson come to mind: they moved from Sydney to Terania Creek just in time to become centrally involved in what became one of the first major forest conflicts in Australia. Without the alternative lifestylers dramatically changing the demographic make-up of some very conservative rural communities, these direct actions may never have happened. These traditions continue at the turn of the millennium under names such as permaculture, championed by Bill Mollison and others.

Then there are those who are closely linked to certain organisations and groups that have been involved in a vast number of campaigns. One is the late Milo Dunphy, who for a long time was involved with the Total Environment Centre in Sydney (Meredith 1999). In Melbourne, Geoff Mosely continues to shape the Australian Conservation Foundation as a councillor, and he served as director through its toughest years. Unionist Jack Mundey tried to drag environmental consciousness into the cities in his Sydney-based green bans in the 1970s. Lessons learned from this urbanisation of green issues are still being read, and will feature much in the movement's future. Benny Zable was (and remains) the personification of the anti-nuclear movement, the 'man in the suit of death', a one-person piece of street theatre at every rally and a silent sentinel to the doomsday clock.

Then there are the places: precious ecosystems or places of conflict that were the reason for all of these battles. They assume their place in a mythical honour roll, pronounced with hushed reverence, now holy symbols of the movement: Great Barrier Reef, Colong Caves, Lake Pedder, Terania, the Franklin, Daintree, Pine Gap, Roxby, Nightcap, Errinundera, Jabiluka.

There were, and always will be, many diverse networks centred around different issues operating within the movement. During this first period, for example, urban campaigns emerged to question existing infrastructure such as the blossoming of freeways in major cities, to promote public transport, and to imagine alternative forms of energy. Despite its diverse manifestations—the sheer number and magnitude of the issues that fall under the environment banner—the environment movement in Australia has been dominated by wilderness issues (Hay and Hayward 1988; Eckersley 1992; Doyle and Kellow 1995).

'Wilderness' covers a vast range of associated subjects, approaches and campaigns. The concept largely addresses the value of the non-human world or what traditional science refers to as 'nature'. Wilderness can be valued instrumentally or intrinsically. Instrumental value is awarded by humans (seen as distinct from wilderness, separate from nature). Warwick Fox refers to one of these perspectives as 'resource preservation'.

He lists nine separate types of argument that are used to justify the value of wilderness in human terms (Fox 1990, pp 154–61). These include arguments pertaining to the life support system; the early warning system; the laboratory argument; the silo argument; the gymnasium argument; the art gallery argument; the cathedral argument; the monument argument; and the psycho genetic argument. It is crucial to understand that these types of justification for preserving wilderness pockets are still anthropocentric (human-centred). It is these resource preservation arguments that have grounded the dominant wilderness perspective in the Australian environment movement.

The other more recent, substantial but subservient tradition—and a philosophically more radical one—accepts that non-human nature has intrinsic value (regardless of human value). Robyn Eckersley (1992, p 49) refers to this as ecocentrism: 'According to this picture of reality, the world is an intrinsically dynamic, interconnected web of relations in which there are no absolute discrete entities and no absolute dividing lines between the living and the non living, the animate and the inanimate, or the human and the non-human.'

There are many subcategories of this position. Fox (1990, p 162) lists four deviations of 'intrinsic value theory': ethical sentientism, biological ethics, ecosystem (Gaian) ethics and cosmic purpose ethics. In this general case, however, humans become part of nature, equal component parts with other species and natural entities. In this world view, at least theoretically, the separation between humans and the rest of nature is banished. Strategically, though, as ecocentric politics has been played out, much of the emphasis of ecocentrists (or deep ecologists) has been less holistic as it has focused on those parts of nature that have been the least disturbed by white Australians: wilderness areas.

In some forums this dominant wilderness focus—whether it be fuelled by anthropocentric resource preservation or ecocentric arguments—has been attacked by environmentalists who are not driven by wilderness arguments, but whose political aspirations lie in other green traditions such as 'resource conservation', 'human welfare ecology' or 'social ecology' (Eckersley 1992, pp 33–47) (see also chapter 4). By conceiving the 'environment' as protecting and valuing wilderness refuges and pockets, wilderness advocates have been slighted for being, on occasion, anti-human, anti-urban and anti-rural. Much of this criticism has portrayed the wilderness networks of the movement as middle-class, wealthy professionals looking to parcel up tracts of nature.

It is always important to remember before we lash out at conflicting radical environmental traditions what the dominant views of the 'environment/nature' are. The unrestrained use ideology portrays non-human nature as either totally expendable or as something that can be transformed, made better, conserved and developed simultaneously. Wilderness-oriented perspectives have also dominated elsewhere, particularly in Scandinavia, North America (most specifically the western regions) and New Zealand. This has occurred, in part, because of a dual reality. First, in these countries, there are still large tracts of relatively 'undeveloped wilderness' (Eckersley 1992, p 70)—unlike in most European countries—and a large proportion of the citizens in countries where 'wilderness' remains are wealthy enough to be able to afford to define the 'environment' symbol in this way (unlike many countries in the 'developing world') (Doyle and McEachern 1998).

Forests dominated wilderness concerns in the early movement in Australia. Even in South Australia—where there are no large, native hardwood forests to speak of—the issue of native vegetation clearance has assumed some importance. Forestry issues have demanded so much attention that some environmentalists interested in other habitats have been reluctant to mention the 'f' word.

Despite the fact that Australia has vast coastal, arid and semi-arid wildernesses, nearly all wilderness conflicts have occurred in temperate, sub-tropical or tropical rainforests. Their profound beauty generates passion, and the fact remains that forests are just so directly under threat. They inhabit sections of our continent that we humans also like to inhabit: the coastal areas where there is reasonable rainfall. It is no wonder we have seen conflict on their behalf. This primacy of the wilderness agenda began early in the movement's development. When compared to movements elsewhere in the world, this is *the* defining characteristic of Australian environmentalism.

Probably the next most dominant collection of campaigns on the national agenda has been those relating to anti-nuclear/anti-uranium mining issues. The anti-nuclear and the wilderness campaigns were characterised by very different notions of power and by conflicting ideas about the movement's relationship with the state. Many of these fundamental political differences can be understood with the aid of two different models of policy analysis: pluralism and structuralism.

PLURALISM, STRUCTURALISM AND THE STATE

Pluralist and structuralist definitions of power inform the movement's relationship with the state and provide an explanation of the underlying

THE FIRST PERIOD: UNRESTRAINED USE • 119

ideology for the movement's campaign strategies and political decisions. Concepts of what constitutes the state are diverse. Particularly in
recent times, the divisions between the state and other parts of both
the private and public spheres have become somewhat confused. Ham
and Hill's (1993, p 23) definition of the state, however, provides us
with a starting point for discussion:

> The state can be defined both in terms of the institutions which make it
> up and the functions these institutions perform. State institutions com-
> prise legislative bodies, including parliamentary assemblies and subordi-
> nate law-making institutions; executive bodies, including governmental
> bureaux and departments of the state; and judicial bodies—principally
> courts of law—with responsibility for enforcing and, through their deci-
> sions, developing the law.

In the classical pluralist view of society, it is assumed that there are
no significant concentrations of power in society: power is diffused and
all citizens have some power resources they can use to achieve their
aims or interests. No group is considered more privileged, and no
group's access to the democratic system is blocked. The important
concept here—in terms of policy formation and implementation—is
that power is equally distributed.

Individual citizens with shared visions, grievances or interests are
free to form organisations ('interest' or 'pressure' groups) to press
their claims in the political process. Sometimes, the state responds to
these demands and some adjustment can be achieved. In the pluralist model, environmental NGOs can indirectly correct minor problems in the existing political system; while larger problems,
including fundamental, systematic flaws remain unacknowledged
and unresolved.[1]

In this model the state is treated as 'neutral', having no particular
'axe to grind'. In pluralist systems green NGOs exist to lobby. They
do not seek public office directly, but through influence provided by
members, public opinion, money and power seek change indirectly. In
addition, there is no notion of informal politics in this schema.
Individuals are perceived as innately apolitical. Only when something
goes wrong do citizens coalesce into formal and legitimate interest
groups and challenge contemporary politics. When the minor hitch
has been acknowledged and addressed by the state, the groups often
disband and, once again, the citizens return to their atomised, individualised and apolitical lives. In many ways, this democratic model
can be seen as the US model of politics: 'anyone can become the
President of the United States'. Anderson (1984, p 51) writes from a
pluralist perspective: 'Public policy, at any given time, will reflect the
interests of dominant groups. As groups gain and lose power and

influence, public policy will be altered in favor of the interests of those gaining influence against the interests of those losing influence.'

Within the confines of this model of politics, environmentalists (and all other interest groups, for that matter) have been depicted as 'outsider' interest groups that battle, with varying degrees of skill and amounts of power, indirectly but legitimately to influence the state: to mend the tear in the fabric of democracy. Figure 7.1 illustrates this model of power distribution, with the 'objective' state at the centre of the political field, with the various interest groups revolving around it, each one attempting to gain the state's attention by increasing 'the decibel reading' of the issues that bring the group into being. The groups seek legislative changes and/or new policies that are favourable to their concerns. The elites within this system are por-trayed as reflecting and responding to the interests of the majority of citizens.

This model of power depicting public policy-making and the rela-tionship between citizens and the state is so dominant in Australia, and in many other industrialised democracies, that it is often not publicly considered as a model or political system, but rather as reality. Consequently, the model is rarely subjected to serious analysis within these societies.

Structuralist analysis presents us with a major challenge and critique of pluralist understandings of power. Structuralism is one example of elite theory, which states that power in all societies is held by a small collection of minority groups run by individuals who have similar social backgrounds. This elite controls the state, big business, the military and other powerful sectors. Gaetano Mosca (1939, p 50), a classical elite theorist, writes:

> In all societies—from societies that are meagrely developed and have barely attained the dawnings of civilisation, down to the most advanced and powerful societies—two classes of people appear—a class that rules and a class that is ruled. The first class, always the less numerous, per-forms all political functions, monopolises power and enjoys the advan-tages that power brings, whereas the second, the more numerous class, is directed and controlled by the first, in a manner that is now more or less legal, now more or less arbitrary and violent.

Table 7.2 outlines four key attributes of elite theories of power distri-bution within societies.

Figure 7.1
A pluralist model of policy-making

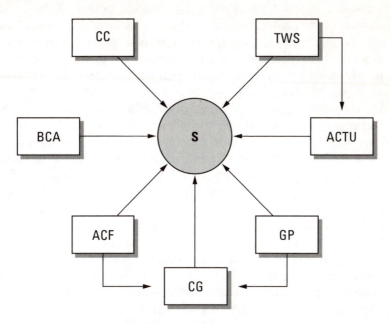

Note: This model shows the state (S) at the centre of the policy-making process, surrounded by many varied interest or pressure groups, such as the Australian Council of Trade Unions (ACTU), the Business Council of Australia (BCA) and environment groups (ACF, Greenpeace, conservation councils and TWS), each acting upon the state and upon each other.

TABLE 7.2
ELITE THEORIES OF POWER DISTRIBUTION

1   Power is held by the few. The majority of people have no input into the policy-making process.

2   The elites are not representative of the population, therefore their policies are not designed to fulfil broad societal goals. These theories also support incremental decision-making, where policy formation is not open to substantial input or change, and basically remains static.

3   Non-elites can slowly move into elite circles, which maintains stability (in that a vision of upward mobility is offered). Before being admitted to elite circles, newcomers must accept the values, guidelines and practices of the ruling regime.

4   Elites influence masses before masses influence elites.

Source: Adapted from Dye and Zeigler 1980.

An incisive contrast between pluralism and elitist models of power relates to the *directionality* of power. As Figure 7.1 (page 121) illustrates, classical pluralist theorists argue that the power of the people surges towards the centre, informing and demanding that their leaders respond to their wishes. Citizens, it is argued, get the governments they deserve. Elite theorists, on the contrary, see power moving predominantly in the opposite direction, with elites shaping the wants and dreams of 'the masses'.

Elite theorists argue that a range of differing forms of social stratification provide the opportunity for elites to dominate. These may be based on gender, race, age, class and many others. 'Class' is just one manner in which groupings can be delineated and ordered. Structural understandings of power are more often than not centred on class-based analysis, and I use structuralism here as just one example of elite theories of the state.

Class-based analyses, originally forged in the writings of Marx, have been roundly savaged by neo-pluralist writers supporting the relatively recent emergence of a US-led world order. This is not surprising, as class-based analysis remains a most effective critique of advanced global capitalism. This is quite separate to the issue of whether Marxist theory can present cohesive and workable societal alternatives. Class-based analysis is also salient in terms of the environment, particularly when nature is so often construed as a series of resources that provide the fuel and raw materials for markets. Neo-Marxists such as Ralph Miliband contend that the state is not neutral; but rather an instrument of the ruling elite, the bourgeoisie (the owners of the means of production) to promote its own interests. Ham and Hill (1993, p 35) describe this position as follows:

> First, there is the similarity in social background between the bourgeoisie and members of the state elite ... Second, there is the power that the bourgeoisie is able to exercise as a pressure group through personal contacts and networks and through associations representing business and industry. Third, there is the constraint placed on the state by the objective power of capital.

The instrumentalist position has been criticised by pluralists on the basis that hard evidence of the existence of cartels of powerful conspirators plotting to disempower the majority of people while shoring up their own nests is hard to find. A far better criticism emanates from other Marxist scholars. One such scholar, Nicos Poulantzas (1973), builds on the work of Miliband but argues that he overemphasises the importance of the shared class background of state officials and the bourgeoisie. At the same time, Poulantzas argues, the objective power of capital is underemphasised, for the state is not just 'a collection of institutions and functions, but a relationship between classes and society' (Ham and Hill 1993, p 37). In this manner, Poulantzas and other structuralists develop a more sophisticated reading of Marxist theory by perceiving political actors as usually constrained by the political and economic order within which they operate. Far from everyone being able to aspire to be President of the USA, let alone Australian head of state under the British monarchy, structuralists would argue that the choice of citizens is limited and largely predetermined. The everyday mechanics of capitalism determines, to a significant extent, what can and cannot be done in the public sphere.

In this model of politics, environmentalists are seen as being even further outside the state than interest groups under pluralism. Whereas pluralist interest groups seek to influence the state, structuralists are understood to be pursuing more fundamental societal change. Within this view, the interests of the state, capital and elites are largely indistinguishable, so that any incremental changes to the laws and legislation that serve the state are perceived as only further enhancing its legitimacy and authority.

Overt and instrumental abuses of power by business interests to hush opposition are most obvious and visible. The dispersal of the state's controlling power, however, is often more subtle than this. Its effects are also felt in the everyday politics of social life and, as social movements remain a vibrant part of their society, the myths of control are also used to suppress dissent within the movements themselves.

What makes this lack of dissent so disturbing is that Western nations such as Australia constantly profess their credentials as free societies, equally open to all people with all points of view. This pluralist model and its antithesis are the focus of the next chapter, as we investigate the styles of politics practised by the movement to combat the repressive, extremist and environmentally degrading ideology of unrestrained use.

NOTES

1 Interestingly, the 'post-materialist' theory explaining the origins of the environment movement (most clearly articulated by Inglehart 1977) fits nicely within the pluralist polity. Environmentalists are usually construed as just an interest group of middle-class, educated professionals pursuing their 'hobby' of higher order fulfilment.

8

THE EARLY MOVEMENT'S RESPONSE TO UNRESTRAINED USE: PLAYING OUTSIDER POLITICS

During the first period of modern environmental activism in Australia, environmentalists were seen as playing outsider politics, that is, operating outside the state. Consequently, environmental politics were largely adversarial: a battle between the state and/or corporations and environmentalists. Government, private industry and the mass media frequently perceived environmentalists as dangerous subversives, directly threatening the ideology of unrestrained use and the machinations of the status quo. Environmentalists were seen as deviant and radical. Negative stereotypes and folk devils were created to illustrate to the 'average Australian' who these mutants were. They were depicted as different in every way, from their clothing to their diet, their ideas to their political activities. At the height of the second Daintree blockade in August 1984 environmentalists were described as follows in the north Queensland press:

> Dirty, smelly, anaemic looking individuals—most looking as if they could do with a good feed, including meat ... dole-bludging lay-abouts and non-conformists whose main aims are to disrupt normal society in the most effective way possible ... a rabble-rousing collection of petty crims and society drop-outs and non-descripts. (McGee 1984)

In these circumstances it took real courage to be an environmentalist. It was not something people took on lightly. As the movement included the more traditional nature conservationists, other more conservative groups, and still other organisations that were involved in 'politics-as-usual', this pigeon-holing of environmentalists as extremists was inaccurate and unfair.

Outsider politics can take two forms: legitimate, formal and reformist on the one hand, and illegitimate, informal and radical on the other (or outside and *way* outside!). Legitimate outsider environmental campaigners pursue politics within the rule of law, attempting to influence established political parties, lobbying for constitutional reform, involving themselves in legal and quasi-adversarial proceedings, pursuing international heritage status for local sites, seeking incremental changes through legislation and politics-as-usual. Brian Martin (1984, p 110) refers to these as 'appeal-to-elites' strategies:

> The assumptions of the appeal-to-elites approach underlie most social action on environmental issues. The basic idea is to convince key decision-makers in government and sometimes industry of the logic and justice behind taking action and making policy to solve environmental problems. Typical appeal-to-elites methods are writing letters to politicians, sending petitions to politicians, making submissions to environmental inquiries, lobbying politicians and government bureaucrats, and writing articles aimed at elite policy-makers. The necessity for adopting such methods based on 'speaking truth to power' is seldom spelled out, but simply assumed.

Martin's final point about unquestioned assumptions is a crucial one. Many of these appeal-to-elites strategies are pursued without sufficient critical gaze, as these political pathways are simply considered obvious and 'natural'. Underlying these assumptions is the myth of a pluralist polity. Pluralist notions of power, and the imagined relationships between dissenters (recast here as interest groups) and the state determine, to a large extent, the selection of 'legitimate' methods and strategies utilised by social movement actors. Martin (1984, p 111) contrasts the methodology of grassroots mobilisation with appeal-to-elites strategies as follows:

> Grassroots methods aim at mobilising 'ordinary people' in all walks of life to promote social change by collectively changing their behaviour. Grassroots methods do not rely on support from elites for achieving their goals. Rather, they strongly encourage participation in a meaningful way by as many people as possible. A typical approach is to organise within a group of some sort—workers, church members, students, parents—at the level of the 'rank and file', by providing information, building networks, fostering development of skills and initiative, and taking action.

Mass-mobilisation methods are more radical than appeal-to-elites ones. They are usually employed by activists who are more conscious and critical of the prevailing models of politics. Grassroots methods are regularly adopted by people who believe that politics-as-usual is insufficient for resolving vast social and environmental problems. Appeals to those in power are often simply appeals to elites with vested interests to forego some advantage for the benefit of others. Consequently, mass-mobilisation techniques are usually used by those who seek more fundamental systemic changes, challenging the continued power of particular elites. This often leads these activists into involving themselves in grassroots, informal politics, and favouring strategies of deep-seated change within longer timeframes. This is 'illegitimate' outsider politics. This approach is supported by elitist/structuralist understandings of power and the state.

On many occasions, there is tension within campaigns over which strategic direction to take. At other times, networks that emerge and then cluster around specific issues can be dominated by certain kinds of strategies and actions. Apart from constituting the two most prominent sets of environmental campaigns in this early era, the wilderness and anti-nuclear campaigns are extremely useful in differentiating between these different modes of outsider politics. The early wilderness campaigns were archetypes of pluralist, legitimate outsider politics, whereas the anti-nuclear activists often upheld structuralist traditions in attempting to effect far broader societal change. These divisions are simply major trends, not rigid categories. Obviously, there were occasions when networks within wilderness campaigns also challenged the politics of the status quo and, similarly, minority networks within anti-nuclear campaigns played pluralist, mainstream politics.

EARLY WILDERNESS CAMPAIGNS

Many commentators attribute the beginnings of a nationally identifiable wilderness movement (or wilderness networks within the broader environment movement) to the campaign to save Lake Pedder in Tasmania's spectacular and densely forested south-west (Hay 1987, pp 4–12; Pybus and Flanagan 1990, p 17). In July 1972 the 'extraordinary lake with a huge fringing beach of white quartzite sand' (Hay 1994, p 5), situated in a national park that bore its name, was drowned by waters of the newly constructed Serpentine Dam, flooded by the now infamous Tasmanian Hydro-Electric Commission. People such as Dick Jones, Olegas Truchanas and Brenda Hean were centrally involved in this issue. Hean and her pilot Max Price died tragically at the height of the controversy in mysterious circumstances.

The strategies for saving this beautiful alpine lake emerged in a rather ad hoc fashion. Despite this and the fact that tactics had to be determined in the heat of battle, many of these strategies were prototypes for those used in campaigns over the next 15 years. For example, an initial ploy involved the now ubiquitous signature drive. In 1967, the greatest number of signatures on any petition in Tasmania's history (up to that date) was achieved when the names of 10,000 citizens were gathered in opposition to the flooding of the lake and its pristine ecosystems. This attempt to influence the Legislative Council (Tasmania's upper house) resulted in the establishment of a select committee of inquiry, which ultimately endorsed the government's decision to build its dam in spite of opposition. A similar campaign attracted 17,500 signatures in mid-1972. Another inquiry was set up after the election of the Whitlam government at national level in 1972: the Lake Pedder Committee of Inquiry (chaired by John Burton). When it published its findings it demanded a moratorium on the flooding of the lake, but lost support from the federal government for proposing that the latter should meet all costs associated with the moratorium, including those incurred by the Hydro-Electric Commission through changes to its plans.

In 1971 the informal Lake Pedder Action Committee (LPAC) had come into being and worked alongside the more formal Tasmanian Conservation Trust. The action committee mobilised a network of many local, state, national and international groups and organisations. Apart from the trust, these included the Australian Union of Students, the national parks associations of NSW and Queensland, the Wildlife Preservation Society of Queensland, the conservation councils of South Australia, Western Australia and the Northern Territory, the Royal Australian Institute of Architects, the National Trust, the ACF, the International Union for the Conservation of Nature (IUCN) and a range of bushwalking clubs (Lake Pedder Committee of Inquiry, 1974). Geoff Holloway (1990, p 39) describes the structure and role of the LPAC:

> The Lake Pedder Action Committee was marked by its lack of inflexible hierarchy and the utilisation of the recreational networks within the bushwalking clubs. The Committee set out to mobilise Australia's first national conservation campaign and certainly led to the greater commitment to 'activist' methods by more and more conservationists.

This network-building was far more than just creating 'feel-good' solidarity. This ambitious and multi-tiered form of national and international networking had not been seen in Australian environmental circles before. Using this network with diverse levels of clout and a range of skills and political leanings, the Lake Pedder Action Committee further diversified and intensified its campaigning strategies from 1971. Today these strategies may not seem remarkable, but before Lake Pedder few of these avenues had been tested in a national wilderness campaign. For example, there were strategies aimed at educating more and more Tasmanians and mainlanders about the beauty and natural wealth of Lake Pedder. In one weekend in January 1971, 1000 people visited the lake, which had previously been regarded as accessible only to experienced bushwalkers. People who made the 20-kilometre trek and 'discovered' the lake for themselves became fierce advocates of preservation. Public meetings were called to further mobilise support, media campaigns were embarked upon,[1] symposiums were held, television programs were developed and aired, and a referendum was called for (Kiernan 1990, p 23).

Mainland politicians were also lobbied, with Prime Minister William McMahon increasingly being pressured to intervene. There were also calls for the labour movement to boycott the building of the dam. Jack Mundey, made famous by the green bans he initiated with the Builders Labourers Federation in inner urban Sydney, called for a 'blue ban' in Hobart Town Hall to save Lake Pedder. The Tasmanian unions, however, were dominated by conservative factions of the Tasmanian Trades and Labour Council, and rejected Mundey's pleas (Kiernan 1990, p 32).

At the international level, the opinions of the 'independent' scientific community were sought through UNESCO. The science of ecology was still relatively new, and many of its arguments of 'inherent wilderness values' and the 'interconnectedness of ecosystems' were being heard in the popular domain for the first time. Another petition was mounted, this time aimed at the scientific community: it garnered 184 signatures and demanded a more thorough, scientific examination of the key issues.

There was some concern about the legal ramifications of flooding a national park. The legal system was used only fleetingly in the Lake Pedder crusade when compared with the campaign at Terania Creek in New South Wales, but legal arguments were introduced into forest wilderness debates here for the first time in Australia. Probably the most widely celebrated aspect of the campaign in relation to Pedder was the formation of the United Tasmania Group (UTG) in April 1971: the world's first green party (Rainbow 1992, pp 321–46). The UTG fought the 1972 state election. The local media were largely won over by the Hydro-Electric Commission and the major parties. Two UTG candidates narrowly missed being elected (Kiernan, p 25). The UTG fought 10 elections between 1972 and 1977, but didn't ever win a seat (Holloway 1990, p 39). Yet the emergence of wilderness networks in electoral politics, in the form of the UTG, broke sufficient ground to enable the formation of green parties elsewhere in Australia.

Electoral politics was also being played in the federal sphere. Apart from the UTG, support was being drummed up for a major political party. In circumstances that would uncannily reverberate 10 years later during the final days of the Franklin Dam campaign, the last throw of the dice for Pedder was tied to a Labor victory in the 1972 federal election. Whitlam had promised that, if elected, the Lake Pedder catastrophe would be averted. Despite the ALP victory, the state government refused to negotiate with the new—and, it must be said, lukewarm—federal government on the issue. The flooding of Pedder ensued.

> A vigil camp had been established on the Pedder dunes to monitor the rising waters and to rescue drowning wildlife. The vigil was kept by Chris Tebutt and others until only the crests of the dunes stood above the flood ... Arriving at the top of the Coronets we were appalled: no more basin of light from the reflecting sands, it was merely a drab grey sea and an arc of treetops ... The place was desolate. Never had the air seemed so full of death. (Kiernan 1990, pp 31–2)

The first major wilderness campaign focused directly on forests was Terania Creek in 1979. As we saw in the previous chapter, newly settled residents classified as 'alternative lifestylers' were joined by other activists to protest against the NSW government's plans to log the area.[2] The environmental activists' tool kit was expanding. New tactics, strategies and techniques were being gathered from a range of different campaigns. As with Pedder, there was much lobbying of political elites. Paul Landa and Neville Wran were key movers and shakers after the NSW Labor Party retained government at the 1978 election. At many different stages during the campaign, deals were struck with Labor, but this was fraught with difficulties, as Labor's position seemed to alter, and could not be relied upon.

Two critical additions to the activists' repertoire emerged at Terania: first, the Terania Creek Inquiry, chaired by Simon Isaacs QC, a retired judge of the Supreme Court of NSW (Taplin 1992, p 156), saw the environment movement fully engaged in legal battles. Although the inquiry was only quasi-adversarial, set up to mimic legal proceedings, it was the precursor of many hours spent dealing with every level of the legal system in years to come.[3] The inquiry found in favour of logging interests, only to be overturned by Premier Neville Wran in a still confusing but much welcomed volte-face.

Even more important, however, was the emergence of non-violent, direct action in Australian environmental politics (see chapter 3). This form of political protest, which involved participants putting their bodies in front of bulldozers and thereby passively breaking the law, became one of the most celebrated strategies of the modern green movement. It became a form of theatre—a passion play—its images now entrenched in the psyches of all Australians. It was a defining moment for modern environmentalism.

The fight for the Franklin, which climaxed in the early 1980s, is the most famous wilderness struggle. As the Franklin issue again centred around plans of the Hydro-Electric Commission to flood another part of the south-west of Tasmania, many networks involved in Pedder were reactivated. Strategies honed a decade earlier were brought into play and bolstered by direct action and legal strategies initiated at Terania. For example, in December 1982, environmentalists staged several blockades at the construction site of the Franklin Dam. Twelve hundred demonstrators were arrested 'and the unprecedented public outcry forced [Prime Minister] Fraser to act', and in the interim the Franklin area was classified as a World Heritage site. Just after Labor won the 1983 federal election, the High Court ruled that the Commonwealth government's external affairs powers 'allowed Federal legislation giving domestic effect to international treaties, such as the World Heritage Convention, to prevail' (Tighe 1992, p 126).

During these early stages of the movement, wilderness campaigns were dominated by appeal-to-elites strategies played out on what was accepted as a pluralist political stage. Legitimate outsider methods included: lobbying and electoral politics in all three of these wilderness campaigns, media saturation, submission and petition writing and recourse to the law. In certain instances, these strategies were successful in achieving clearly defined short-term goals, such as the cessation of logging at Terania and the halting of construction of the Franklin dam.

There might seem to be some exceptions. Many of the ideals that framed wilderness debates can only be construed as radical and far-reaching. By questioning, among other things, the fundamental anthropocentric tenets of unrestrained use, wilderness activists suggested a revolutionary change in values. However, although these wilderness values and interests were understood by the state (and the environmentalists themselves) as occupying a position outside the state, the political processes used by the greens were aimed at incrementally reforming the existing order of things.

Another exception may relate to the mass blockades at Terania and the Franklin. Surely these must be counted as grassroots, mass-mobilisation strategies, more in line with structuralist opposition? Martin (1984, p 116) disagrees:

> Civil disobedience can be used for several purposes. It can be: a method for involving people in meaningful experience in challenging unjust laws or actions, a way of demonstrating to others the depth of commitment felt by a group about an issue, a means to obtain publicity and apply pressure on politicians. The Tasmanian blockade was all of those things ... Yet for the TWS organisers, the blockade was mainly used to obtain publicity and thus to apply pressure on national politicians, it was not seen as part of a long-term strategy involving grassroots involvement in non-violent action.

Regardless of what appears to be a plethora of campaigning nodes, the battle for the Franklin was won by environmentalists largely due 'to the structures of Australian federalism and the power of the Sydney and Melbourne media to transport the dispute beyond Tasmania' (Hay 1987, pp 7–8). Although mass blockades constituted a thrilling piece of theatre that was attractive to media obsessed with colour and movement, ultimately this media focus was used to achieve changes in government, important legislation, and constitutional law.

Brian Martin (1984, p 116) is critical of the short-term goals of The Wilderness Society and its appeal-to-elites strategies. He sees opportunities missed for longer term and more fundamental political change:

> Nevertheless, the TWS made little headway in promoting wider awareness of programmes for bureaucratic reform or of alternative methods for decision-making about wilderness. The main emphasis throughout the campaign was on saving the Franklin within the context of present political structures, by a change in policy at the top, rather than a restructuring of political institutions.

Peter Hay (1987, p 8) directly takes issue with Martin's statements by defending TWS's strategists:

> The Wilderness Society has been criticised for not adopting strategies more likely to result in a shift in positions, but criticism on this ground is difficult to take seriously. From the time Lowe lost his premiership the issue could have been won only if a means was found whereby Canberra could override Hobart—for Tasmania's decision-makers were well beyond persuasion. That the campaign could have been used to effect penitence in such a bastion of the dominant ideology as the [Hydro-Electric Commission], which Martin stops only just short of arguing, is clearly ludicrous.

This intellectual altercation between two movement academics raises important issues relating to appeal-to-elites strategies. It frames the two divergent positions of structuralism and pluralism. Martin is arguing from an implicit elitist/structuralist position, seeing 'real' change as having to be deep-seated and long-term, while Hay, in this analysis, argues from within a pluralist political framework, which is oriented to the achievement of specific, short-term goals, working within the constraints of the 'dominant ideology' of the state in Tasmania.

Worse than opportunities lost, however, Martin also perceives despairing outcomes resulting from these strategies. In dealing almost exclusively with elites, there is also a need to rely more and more on professional movement lobbyists/elites. This resulted in increased centralisation of power in the anti-dam movement, with a concomitant loss in participatory democracy and chances of grassroots transformation. The administrative and organisational trends that emerged in these early wilderness campaigns—and most specifically during the Franklin issue—were the harbinger of vast changes to the operational philosophy of the movement. These changes ultimately led to the more corporatist style of environmental politics described in the next two chapters.

There have been many wilderness campaigns, and all have borrowed from, and contributed to, the range of tactics developed in the forests of the Franklin, Terania, Pedder and beyond. The longest running forest wilderness campaign emerged directly after the fight for the Franklin. The battle for the Daintree and the wet tropical forests of north Queensland was a central, national environmental campaign that ran for a decade (see chapters 2, 3 and 10). Wilderness campaigns continue as the major and dominant strand in the politics of the environment in Australia.

EARLY ANTI-NUCLEAR CAMPAIGNS

Anti-war campaigners have been extremely vocal in Australian society since the intervention of the British in Sudan in 1885 (Holloway 1990, p 41). Specifically since 1917, they have often been branded as 'enemies of the state' and 'communists' (Vallentine 1990, p 4). Despite some membership crossover, the peace movement was quite distinct from the environment movement until the mid-to-late 1970s. After this period, it is fair to say that the two movements fused, with the anti-nuclear networks increasingly being seen as part of the environment movement. During this period, the peace movement moved its fundamental focus away from peace and anti-militarism to Australia's involvement in the nuclear power cycle. Australia held 15 to 20 per cent of the world's uranium reserves and was thus in a position to affect the future of nuclear development. In their short history of the Australian Peace Movement, Summy and Saunders (1986, p 45) argue that this transition from traditional peace and anti-war concerns to anti-uranium interests broadened the list of movement goals (and its membership) immensely:

> Their objections were not confined to the link between nuclear-weapons proliferation and the expansion of the nuclear power industry, but were based on a host of other factors as well. These included the environmental hazards of nuclear-waste disposal, reactor accidents, and releases of radioactivity at various stages of the nuclear-fuel cycle. Other problems raised included the social and political threats to a society from terrorist use of nuclear materials, from reduction in civil liberties, and from the centralization of political and economic controls in the hands of financial bureaucratic elites.

Early 'anti-nukes' networks were motivated by revolutionary goals and used battle plans based on instrumentalist and structuralist world views. Anti-uranium campaigning reflected an organised effort towards transforming capitalist society. It also challenged an over-reliance on technology and its experts, and investigated new ways of conserving fossil fuels and developing alternative forms of energy. The formative days of the anti-nuclear movement were thus deeply immersed in social change philosophy.

For example, in 1975, Friends of the Earth (see chapter 5) became increasingly vocal and active in their anti-nuclear campaign. In fact, FoE's anti-nuclear campaign was the organisation's 'coming-of-age' in Australian green politics. FoE was involved in battles across a range of fronts. First, it responded to the 'Ranger Inquiry' through submissions and through the media. The inquiry was set up by the Fraser government under Mr Justice Fox, and included in its terms of reference investigations into most elements of the nuclear fuel cycle and the nuclear industry across the globe. FoE, however, was fiercely and consciously committed to structuralist challenges to the industrial/militarist state, and was committed to grassroots activities. Early issues of its magazine, *Chain Reaction*, are riddled with references to the underlying philosophy of these tactics. The Summer 1975 (p 3) edition reads:

> We held our first national meeting (and pillow fight) with nine delegates from four states from December 9–13, 1974 on a proposed nuclear reactor site at beautiful French Island, Western Port Bay (Victoria). The politics of the middle-class and of FoE were much debated. One delegate believed that all FoE has done has been to continue in the tradition of middle-class politics of manipulation. It was emphasised that what must be changed are the attitudes of the population at large, and that it is all too easy to slip into a mode of action of the lobbying of different elite strata—trade union leaders, politicians, students, other environmentalists … It was agreed that FoE's objectives are to consolidate the alleged conservation power base in the community, and to crystallize the vague environmental awareness into specific demands and action.

FoE also used other grassroots strategies, including street theatre, anti-nuclear bike rides to Canberra, petition drives, alternative energy fairs, community development through local participatory politics, and mass demonstrations and direct actions. In 1976 and 1977 up to 50,000 people gathered in mass rallies in major Australian cities to voice their opposition, and in June 1977 anti-uranium demonstrators directly tried to stop the ship *ACT 6* from exporting yellowcake from the Lucas Heights reactor: 40 arrests ensued. Although these mobilisation tactics were local, they were also global. FoE was, and still is, an international organisation with transnational networks. Although there were attempts to reach out to the international community in the early

Australian wilderness campaigns, the anti-nuclear movement was rooted in the international community in the first place. Similar mass protests and other forms of opposition were occurring across the industrialised world. Table 8.1 lists some of these events.

TABLE 8.1
MASS PROTESTS AGAINST URANIUM IN THE MID-TO-LATE 1970s

1974	Death of Karen Silkwood, a worker at the Kerr-McGee plutonium plant, Oklahoma, in mysterious circumstances;
1975	Mass protests against the proposal to construct a nuclear power station at Kaiseraugst, Switzerland;
1976	Mass protests against the proposals to construct nuclear power stations at Brokdorf and at Whyl, Federal Republic of Germany;
1977	Mass protests against the proposals to develop nuclear power installations at Grohnde, Kalkar, and Brokdorf in the Federal Republic of Germany and at Creys-Malville in France, and publication of an influential book by Amory Lovins, *Soft Energy Paths: Towards a Durable Peace*;
1978	Mass protests against the nuclear power reprocessing plant at Gorleben, Federal Republic of Germany;
1979	In protests against government policies on nuclear energy around 100,000 people assemble in Hanover and 150,000 in Bonn, Federal Republic of Germany; the accident at the Three Mile Island nuclear power plant in Harrisburg, Pennsylvania; 75, 000 people gather in Washington DC to take part in demonstration against nuclear power.

Source: Adapted from Papadakis 1998, pp xv–xvi.

FoE's action philosophy was the forerunner of the widely used maxim 'Think global, act local'. FoE was joined by a network of groups, organisations and political parties (the Australia Party and the Communist Party of Australia), socialists, anarchists, feminists, human rights supporters, pacifists, Aborigines and traditional nature conservationists. By 1976 Campaign Against Nuclear Power, Campaign Against Nuclear Energy, and Movement Against Uranium Mining had formed. Mainstream environment organisations such as the ACF and TWS were also in the network.

This emphasis on mass-mobilisation strategies, including direct actions, might justify what was to become the most influential decision by anti-nuclear campaigners in this early period: to align the environment movement with the labour movement. Clearly, organising at the shop-floor level of the union movement could be justified within this strategy. In 1976, at Mary Kathleen mine, a 24-hour strike was called by the Australian Railways Union on the issue of nuclear hazards. In

1997, another 24-hour strike was called on the Melbourne wharves by the Waterside Workers Union, with a ship carrying a load of yellow-cake, the *Columbus Australia*, blackbanned. This coincided with anti-nuclear protests at the wharf. In 1977 the ACTU adopted anti-nuclear policies, with the demand for a referendum on uranium mining (Green 1998, p 14). In 1981 the Seamen's Union also attempted to prevent uranium exports from Darwin, but failed under threats of union dereg-istration by the Fraser government.

But the labour movement itself was in the midst of major power struggles, which resulted in right-wing factions assuming authority. Right-wing unions such as the Australian Workers' Union actively white-anted this anti-uranium stance. Furthermore, the then president of the ACTU, Bob Hawke—later dubbed 'Yellow Cake Bob'—worked assiduously up until the early 1980s to promote uranium mining, despite the majority view. Ultimately, the decision to join forces with the labour movement saw major challenges emerge to anti-nuclear campaigners' philosophy of grassroots mobilisation, and a move towards tactics more reminiscent of the appeal-to-elites strategies of their wilderness counterparts. The environment movement was delivered—via the labour movement—to the Labor Party and the merry-go-round of electoral politics.

The Labor Party and the labour movement were quite closely con-nected during this period. Until the mid-1970s, the ALP was a passive supporter of the nuclear industry. Uranium mining had taken place in Rum Jungle in the Northern Territory and Mary Kathleen in Queensland between 1944 and 1963. It was not until 1974, however, that mining recommenced at the latter site: this was during the term of the Whitlam government. After it was dismissed from office in 1975, the Labor Party, influenced by its members in the labour movement, and convinced of electoral advantage, decided to change its position, and actively campaign against uranium mining. This stance was further strengthened at the 1977 ALP national conference in Perth. In part, the Melbourne Waterside Workers' strike in July of that year acted as a catalyst for this. In response, networks within the anti-uranium move-ment began to concentrate resources into electoral strategies, cam-paigning for the ALP in marginal seats in the 1977 and 1980 federal elections. The answer to the movement's problems no longer seemed to be rooted in the struggle against the structural constraints of capi-talist/militarist societies, but rather with the successful election of the ALP. There were always grassroots strategists who chose to renege on any commitment to Labor, but these networks—and their philosophies of action—ceased to dominate.

Labor lost both those elections, and so did the anti-nuclear move-ment through its affiliation. In 1982, the ALP watered down its anti-uranium stance (Holloway 1990, p 42) and in 1983, when Hawke and Labor came to power, the party soon adopted the 'three mines' policy: mining could continue at Roxby Downs in South Australia (see chapter 12), and Ranger and Nabarlek in the Northern Territory. Since that time, the Labor Party has progressively bowed to uranium mining interests. The Franklin Dam 'softener' eased the pressure generated by the uranium volte-face under Hawke.

Following Labor's decision there was a momentary return to prominence of mass-mobilisation tactics. Introduced in 1982, 'Palm Sunday rallies' peaked in 1984 and 1985, with over 250,000 people attending in Australian cities. Deterred from the major parties, but still focused on electoral politics, the anti-nuclear networks of the environment movement did what Lake Pedder activists had done a decade earlier: they created their own political party, the Nuclear Disarmament Party (NDP), formed in the foyer of the Lakeside Hotel in Canberra, the venue for the 1984 Labor Party conference.

Not all electoral strategies are appeals to elites. Although accepting the legitimacy of parliaments, the NDP initially practised its own form of party politics in a bid to educate the Australian public about the dangers of the nuclear fuel cycle and Australia's part in the nuclear arms race. It also focused on grassroots participation and empowerment: political *process* became all-important. In the 1984 federal election, the NDP polled 6.8 per cent of primary votes, and West Australian Jo Vallentine (1989, p 62) was elected to the Senate. She reminisces:

> The campaign in WA was a very empowering experience. People came in and offered to work and worked very successfully at community level, it being mostly women who did the work, and often women with small children. Sometimes our office was a nightmare of kids' trolleys. But this was also a fantastic reminder of what and who we were all doing it for ... we operated a committee for the length of the campaign without any particular structure. It was non-hierarchical, whoever came along to the meetings being involved in the decision-making.

Others involved in the NDP wanted parliamentary power, and were prepared to play traditional electoral politics in order to get it. Interestingly, it was only in Western Australia, in the branch most committed to participatory politics, that the party achieved electoral success. The NDP waned quickly. Much has been blamed on the Socialist Workers' Party (SWP). Due to over-exuberance on the part of the NDP's founders, members of other political parties were not totally excluded from its ranks. The SWP saw the new party as pursuing similar objectives to its own and promptly moved to dominate proceedings within the NDP. Consequently, as Vallentine (1989, p 63) states: 'They were at our first national conference in numbers out of all proportion to their membership in the overall organisation.' There were other problems within the NDP, however, which also contributed to its demise. In the opinion of some NDP members, Michael Denborough, the founder, had become possessive of the party, refusing, for example, to disclose the source of $30,000 in campaign donations. Also, large parts of the anti-nuclear movement, which had once been based on grassroots mobilisation, now had an administration so centralised that all members of the national executive were from Canberra! At the national conference held in Melbourne after the 1984 federal election, Vallentine walked out in disgust along with other key members Peter Garrett, Gillian Fisher and Jean Meltzer. This occurred before Vallentine had even taken up her position in Parliament.

Environmentalists fighting anti-nuclear campaigns made more successful links with indigenous people's interests than any forged previously by predominantly white activist groups. For example, in July 1975, 14 interstate environmental representatives (10 of whom were from FoE) conducted a fact-finding tour of the Northern Territory, visiting the Ranger mine site with members of the Territory's Environment Council and meeting with the local Oenpelli people (Hutton 1999, p 138). Saunders and Summy (1986, p 45) contend:

> Finally, it was asserted that uranium mining would have disastrous impact on the Aborigines (especially those in the Northern Territory) whose sacred sites and land rights would be violated and their culture generally destroyed—as any 'white invasion' brought with it alcoholism, disease, sexual exploitation, and the breakdown of traditional relationships.

Although there had been earlier attempts by wilderness networks to employ Aboriginal arguments to add weight to their own, these attempts were sometimes tokenistic and exploitative. The anti-nuclear activists did not have to carry this baggage, so a tradition of good, effective relationships between anti-uranium and Aboriginal networks was built during this period, giving a basis for other genuine and prosperous alliances.

The anti-nuclear campaigners within the early environment movement were committed to an ethos of radical social change, which manifested itself as grassroots mass mobilisation, centred around local and participatory forms of politics. This philosophy was informed by elite theories of the state, and was often played out in the rhetoric of class-based struggle: structuralism and instrumentalism. Other discourses used to describe elite-based power also emerged in the anti-nuclear movement's literature: for example, parts of the women's movement perceived nuclear weaponry and its associated industry as resulting from patriarchal power, while anarchists saw a hierarchy of a non-specific kind oppressing the masses in the interests of an elite.

These strong traditions became little more than a continuous undercurrent as more and more movement activity was channelled into labour-based and electoral strategies. These appeal-to-elites methods, more reminiscent of mainstream, pluralist politics, have led to incremental gains (such as the three mines policy), but also to the dramatic weakening of the anti-nuclear movement after the election of the Labor Party to federal government in 1983. Importantly, however, the grassroots philosophy survived the onslaught of politics-as-usual during this period. This continuity in tradition partly explains the successful re-emergence of grassroots mobilisation and protest groups in the 1990s, such as the Jabiluka Action Group (JAG). In chapter 12, in revisiting the anti-uranium campaign of the 1990s, we investigate different environment movement strategies that have surfaced since this first period. Like the wilderness campaigners, the anti-nuclear activists just do not seem to go away.

As a result of playing outsider politics—in all its forms—the environment movement created the template for many of its social movement actions in the future. In structuralist terms, it developed widespread environmental concern, which challenged the dominant ideology of unrestrained use and the elites who espoused it. Furthermore, it forged the participatory politics of a skilful mass movement that still remains. Using a pluralist model of power, it achieved important legislative reforms, particularly in the 1970s. In addition, many of these laws provided the basis for the Commonwealth to participate more fully in environmental policy-making, previously the domain of the states. Some landmark legislation includes: *Environment Protection (Impact of Proposals) Act 1974*, *Australian National Parks and Wildlife Conservation Act 1975*, *Great Barrier Reef Marine Park Act 1975*, and the *Australian Heritage Commission Act 1975*.

In the wilderness campaigns, pluralist conceptual frameworks dominated from the beginning, and a similar scenario engulfed the anti-nuclear campaigners in the late 1970s and early 1980s. One major result was that wilderness and anti-nuclear eco-activists increasingly sided with the ALP, gambling with the support and interests of their broader membership to support the party at a series of elections. Some commentators argue that this cooption by the Labor Party weakened the movement before it had reached its maturity, and that this cooption led to a decade-long decline in the movement's effectiveness (Green 1998, p 14). With the election of the ALP in 1983, the Franklin had been 'saved', and uranium mining's potential had been 'reduced'.

Yet as the ALP managed to weave much of the movement's 'legitimate dissent' into governmental institutions, a strange thing happened: the 'outsider' groups no longer remained on the outside. Those who had played legitimate politics were welcomed into the citadel, while those outsiders with a structuralist and theoretical bent found themselves excluded from the political process by mainstream environmentalists and the state. As movement activists increasingly concentrated their efforts on the appeal-to-elites strategy, they began to take on the characteristics of the state: they became, on many occasions, indistinguishable from the state. So began the second period, characterised by sustainable use and insider politics.

NOTES

1   Holloway (1990) argues that a media-oriented pressure group that formed following the 'failure' of the Pedder campaign—the South-West Tasmania Action Committee—provided 'the organisational foundation' for the creation of the Tasmanian Wilderness Society, now The Wilderness Society.

2   Twenty years later the forests still unified many disparate environmental networks, groups and organisations. In 1995, some of the largest gatherings of people since the Vietnam moratorium gathered and marched in most Australian cities to protest against the Keating government's failure to protect the remaining native hardwoods.

3   For excellent coverage of the Terania Creek Inquiry, see Taplin (1992), who quotes evidence given at the inquiry to illustrate how adversarial systems have profound difficulties in dealing with scientific and environmental evidence.

9

THE SECOND PERIOD: SUSTAINABLE USE

Not long after the arrival of Robert James Lee Hawke at the Lodge, the politics of dissent in Australia had been devastated. So much had been gambled on ALP electoral victories in 1983 and 1984, that many within the movement felt very little had been gained. As far as wilderness campaigns were concerned, from 1984 to the end of the decade, the major campaign to conserve the wet tropical forests (including the Daintree) dominated the national agenda. The Tasmanian Wilderness Society, one of the few winners under Hawke, changed its title to The Wilderness Society, as we have seen, and jumped into the Daintree campaign with considerable enthusiasm. The battle for the wet tropics was, however, a long, hard-fought slog, and energy quickly dissipated. After the 'unsuccessful' second Daintree blockade in August 1984, many environmentalists referred to the wet tropics campaign as a 'lost' campaign. In 1985 the ACF did not renegotiate the terms of a contract for its wet tropics officer stationed in the north. The wet tropics slipped from its position on top of the organisation's agenda. Nineteen eighty-five became the ACF's 'Year of Antarctica'.

Apart from brief and energetic direct actions at Roxby Downs in 1984–85, the anti-nuclear campaigners entered a term of quiescence and severe despondency due to Labor's betrayal. Other campaigns formed, surged and dissipated. The environment movement went into a period of intense introspection during the mid-1980s as a result. Some commentators had written the movement off as 'just a fashion' at this stage, and many mooted its imminent demise. This hiatus occurred for three reasons. First, and most obviously, Hawke and the

Labor Party had managed to channel the movement's energy into ALP electoral politics. When the ALP was assured of its grasp, it did little to further environmental causes between elections. I have described this situation elsewhere:

> ... the ALP Government still felt assured of the environment movement's electoral support after the Franklin victory. The Labor Party would wait until some time in the future when something new could be gained, some bargain struck which would ensure additional support. Such a compact would be struck three years later, in the final days of the [wet tropics] campaign. (Doyle forthcoming)

Second, the state and the general citizenry were growing more accustomed to some of the environment movement's outsider tactics. Pedder and the campaign to save the Great Barrier Reef were now some 15 years in the misty past. Strategies and tactics were becoming more predictable and less shocking. Third, with the inclusion of environmental protest through pluralist, appeals-to-elite politics, the state increasingly found itself in a more manageable position.

Towards the end of the decade, just when obituaries were being written for the movement, there was a huge surge of interest in aspects of environmentalism across the globe. People flocked to join more mainstream environmental NGOs, particularly in the nations of the North, and in 1989 membership numbers in organisations such as Greenpeace and the ACF peaked. There were numerous catalysts for this charge of interest, and some of these events fell outside the domain of domestic politics. In 1987 the United Nations' Brundtland Commission released its report, declaring triumphantly that the legitimate and rational needs of the environment movement were valid and, more importantly, could be incorporated into business and government projects without unnecessary conflict. Indeed, it was here that we initially understood that 'environment was good for business, and business was good for the environment': this concept was central to the ideology of sustainable use, often referred to as sustainable development. Let us now investigate the ideological framework of sustainable development, and some of the ways in which it has been championed by its greatest advocate in Australia: big business.

THE IDEOLOGY OF SUSTAINABLE USE

Growth is the engine of change and the friend of the environment.

—former US President George Bush

A number of writers have examined the notion of sustainable development.[1] During the 1980s and 1990s it became a 'buzzword' and inevitably, 'under such circumstances, it lacks uniform definition and substance' (Simon 1989, p 42). There is no monolithic view of this modern symbol, as there is no uniform perception as to what constitutes the symbol 'environment'. Like other green symbols, it has been used to advocate radical and incremental environmental change. Furthermore, it has been used by business to undermine and attack environmentalists. Its definition is both amorphous and ambiguous. Indeed the continued currency of sustainable use and development is largely due 'to the way it can be used to support these varying agendas' (Redclift 1994, p 17).

While these diverse interpretations are valid, a dominant reading of sustainable development is the one that has been sold by a dispersed array of business interests and advanced industrial nation-states. Beder (1993, p 8) writes:

> Sustainable development is not about giving priority to environmental concerns, it is about incorporating environmental assets into the economic system to ensure the sustainability of the economic system. Sustainable development encompasses the idea that the loss of environmental amenity can be substituted for by wealth creation, that putting a price on the environment will help us protect it unless degrading it is more profitable, that the 'free' market is the best way of allocating environmental resources, that businesses should base their decisions about polluting behaviour on economic considerations and the quest for profit, that economic growth is necessary for environmental protection and therefore should take priority over it.

As Beder points out, sustainable development has managed to place the economic imperative firmly brfore the ecological. Part of its appeal is that environmental goals can be achieved alongside economic goals.

'Nature' is often socially constructed and there are also dimensions of existence that may reasonably lie outside human control (and even understanding). The balance between the advocacy of human agency and the recognition of an 'other' force is a subtle one. But one good thing about the Club of Rome and other neo-Malthusian Earth astronauts and sailors (the Earth was portrayed in terms of spaceships and lifeboats) was that they were explicit in their belief that the Earth was finite. They directly challenged the ideology of unrestrained use. Despite their failure to understand the social mechanics of life—and their denial of such indicators as race, class and gender—their 'limits to growth' arguments were successfully used, on occasions, to challenge the dominant Enlightenment ideal of progress, which could only be ultimately sustained by pursuing industrial and technological growth wherever and whenever, at all costs. During the first period of the modern environment movement some of these doomsday, limits-to-growth arguments were successfully utilised to attack certain development interests that were involved in widespread environmental degradation.

Unlike the neo-Malthusians, however, the sustainable development view throws out the limits-to-growth arguments (Dryzek 1997, pp 123–36). While there may still be limited biota, sustainable development, the key gospel of Agenda 21, tells us that through increased efficiency and effectiveness in production, these biophysical limits can be bypassed.[2] Thus sustainable development removes humanity from nature in a way that is more complete than the unrestrained exploitation and expansionism that has dominated the planet since the industrial revolution, until its challenge from the doomsayers in the 1960s. What makes it more dangerous is its lack of explicitness and its Orwellian double-speak. While the unrestrained exploitation of the earlier Enlightenment project placed humanity (Man) apart from and in aggressive opposition to an instrumental nature, sustainable development simply does not recognise it: it leaves 'nature' off the agenda. And whereas the neo-Malthusians questioned absolute limits, sustainable development denies their existence.

Interestingly, although there are some important differences, the sustainable development view is not dissimilar to the traditional Marxist line. In a socialist transition, the proletariat were to gain control over resources and their own labour. After this transition, by overturning capitalism's inefficiencies and waste, socialism 'could spur the growth process: outdoing capitalism at its own game' (Bell 1987, p 11).

Sustainable development constructs all environmental 'problems' as efficiency issues, which have to be 'managed' more effectively. These management issues include being more technologically, economically, organisationally, educationally and politically efficient. One of the associated concepts that has emerged as part of this ideology is multiple-use planning. Environmental managers using this model argue that most environmental processes are reversible, and that all sectors of society can pursue diverse goals for resource use without unnecessary conflict. All that it takes to succeed is that conflicting sectors produce end positions based on compromise and negotiation. There is no environmental 'crisis' or series of crises within the sustainable development, multiple-use equations, as they still hinge on the belief that humans are external to nature, and that nature is simply a collection of instrumental resources for human use. Piecemeal change is all that is ever necessary. So minor changes to laws and legislation are all that are usually deemed appropriate.

When people are seen as part of the sustainable development equation they are seen as an 'environmental security' issue. People are still not part of the environment, they are simply users or, in the case of the poor, degraders. Some sustainable development economists argue that environmental security leads to national security, which leads to economic growth. The concept of sustainable development emerged from United Nations negotiations during Brundtland, and was later championed by the 1992 Rio de Janeiro Earth Summit and, even more recently, the Earth Charter (see chapter 12). Many of those preaching the ideology of sustainable use present global contexts, have global aspirations and, in turn, the ideology is shaped by global concerns. For example, this sustainable development view, generated largely by the more affluent nations, still contends that the poverty of the South has caused and continues to create environmental degradation. In 1990, the United Nations Human Development Report argued that poverty is one of the greatest threats to the environment and in 1993, the International Monetary Fund (IMF) announced: 'Poverty and the environment are linked in that the poor are more likely to resort to activities that can degrade the environment' (IMF cited in Broad 1994, p 815). There are two key problems with this line of argument. First, all poor people are regarded in a homogenous fashion. This blanket classification is unsatisfactory at best, and derogatory at worst. Second, some sustainable development theorists fail to weigh up the costs of advanced industrialism on a global scale, not just within the boundaries of nation-states. Issues of overconsumption in the North—and by the North—cannot be underestimated. The USA and Japan together represent 40 per cent of the world's gross national product (Imura 1994). Furthermore, the power of transnational corporations and their capacity to degrade environments are not usually addressed within the discourse of sustainable use and development.

Despite this nonsense, 'environment-degrading poor people' in the South remain a grave concern to the North, now advocating global ecology and, as a consequence, seeing itself as inevitably having to share the Earth's essential survival mechanisms with the South. The North now portrays the major problems of the South as species extinction, global climate change, desertification, the international shortage of fresh water, and overpopulation. Needless to say, these are not issues high on the environmental agenda as defined by most people living in the Third World. Other issues of more immediate survival dominate. In a provocative book entitled *Tears of the Crocodile*, Moyo et al. (1993, p 5) argue that the developed world has managed to divest itself of responsibility to the global environment by moving the arena 'away from people and onto things, forces':

> In short the developing world for the first time, is being asked to be an equal partner in a world-wide endeavour precisely because the emphasis has shifted away from the needs of the poor. By advancing an environmental agenda the North has once more concentrated on its own interests and has called them globalism.

Central to increasing this environmental efficiency and environmental security is the 'unfettering' of the market place from the bothersome leftover flotsam and jetsam of 'planned economies'. This entails the globalisation of both radical libertarian free markets and American pluralist/corporatist political models of power distribution. Redclift (1987, p 4) writes:

> Sustainable development, if it is to be an alternative to unsustainable development, should imply a break with the linear model of growth and accumulation that ultimately serves to undermine the planet's life support systems. Development is too closely associated in our minds with what has occurred in western capitalist societies in the past, and a handful of peripheral capitalist societies today.

FREE-MARKET ENVIRONMENTALISM

Adam Smith's invisible hand can have a green thumb.

—US President Bill Clinton

In chapter 4 we briefly discussed the right-wing radical libertarian roots of free market environmentalism. Free-market 'green' economics lies at the heart of sustainable development. Whereas once a liberal democratic state was valued to monitor the excesses of an antisocial minority (such as poor and irresponsible business practices), now the radical libertarian approach dictates that such government 'interference' is actually degrading to the environment, and that good business is good for the environment. Due to this rationale, much environmental legislation and national and state-based environmental protection agencies have been gutted in recent times. Furthermore, many of these agencies have changed their focus from monitoring industrial excesses to helping business jump through the regulatory hoops.

There are no grounds within this ideology for public health care, support for the homeless and the underprivileged, provision of non-profit-inspired international aid, or protection of the environment for purposes other than 'best practice management'. What was once society is now portrayed as an amalgam of individuals—or rather investors and consumers—each pursuing their dispersed and disparate interests in the market place. All of the public losses are justified by arguments of private gains.

As I argued in chapter 4, the classical liberal philosophy, on the other hand, assumes the existence of an antisocial minority (in this context, one that is environmentally degrading), while the majority of citizens will behave responsibly. Consequently, liberal political theorists have long argued that governments have a role in monitoring publicly owned resources, by deciding who gains access and in what circumstances. Their radical libertarian 'free market' counterparts argue, however, that the problem with this theory is public ownership itself. They contend that the resources must be made available to private owner-ship, and its protective boundaries must be removed. All externalities must now be included in the economic, private ownership equation. Free-market economists believe that this is a most positive step. Forests, for example, can now be 'valued' for more than just their wood. Other values can now be included in the equation, including wilderness and whole ecosystem products, with projected future values also added. The negatives, however, far outweigh the more immedi-ately recognisable 'positives'. All values are now economic, nothing is considered to be outside the market: the Earth ceases to be a place for all, both human and non-human, but becomes a series of privately owned commodities. This unfettered system returns us to the barbar-ic, inequitable and misunderstood economics of Adam's Smith's 'free hand' of the market. Through pursuing these approaches, the gap between the poor and the more affluent is getting greater every year. This gap is not imagined. Even the radical libertarians acknowledge it. For example, in July 1996, the United Nations Development Program (UNDP), another Brundtland-style international agency selling sus-tainable development worldwide, began its 'sustainable human devel-opment' report with the following words: 'The global gap between the rich and the poor is widening every day ... If present trends continue, economic disparities between industrial and developing nations will move from inequitable to inhuman.'

But what do the sustainable developers and Agenda 21 disciples advocate in a bid to attack this problem: more economic growth! The report continues: 'More economic growth will be needed to advance human development, particularly for those whom growth has failed in the past' (UNDP 1996). In short, this increased competition will lead to more economic growth. On this occasion, however, growth will be further unfettered by dissolution of international trade barriers, using such Northern devices as the World Trade Organisation (WTO), the International Monetary Fund (IMF), the World Bank, and the Multilateral Agreement on Investment (MAI) (see conclusion for a dis-cussion of the MAI and its antecedents).

Much of this radical libertarian economic debate has led to the pri-vatisation of social responsibility. In 1992, a World Bank report argued that the transference of state industries and services to private owner-ship '*in itself* improves economic efficiency' (World Bank, cited in Martin 1993, p 139). Martin argues that by developing econometric models to justify their beliefs, the IMF and the World Bank 'entirely exclude the testimony of the people affected, reducing their experience to algebraic gobbledegook' (Martin 1993, p 144). Martin goes on to list several instances of environmental degradation in the South due to these Northern radical libertarian interpretations. These include the loosening of controls on foreign logging companies to promote timber exports and thus debt repayment resulting in widespread forest destruction in South America and Asia, and the release of workers from state-owned companies, leading to massive migration into wilderness areas to establish subsistence landholdings in Bolivia and Brazil (Martin 1993, pp 158–9).

Globalised and unfettered growth is only part of the picture framed by Agenda 21 and sustainable development. All parts of the world must also operate on the same political system: American-style pluralism.

PLURALIST DEMOCRACY

The pluralist model of politics is promoted as part of the sustainable development package. In fact, as advocates of sustainable development argue that free markets are essential to its delivery, so too they contend that pluralist democracy is the only political system that can foster it. Much debate about sustainable development hinges on a country's ability to produce political conditions where people can help themselves (as if they were not already). This is referred to as the construction of 'civil society'. It is argued, for example, that once the poor can help themselves, become more environmentally active, then global environmental degradation will cease.

An excellent example is the recent development in Australia of the Centre for Democratic Institutions (CDI). On 27 August 1997 the Minister for Foreign Affairs established the CDI through AusAID 'to assist the strengthening of democratic institutions in developing countries'. One country to 'qualify' for support was, not surprisingly, Indonesia. In this specific case Indonesians are to be trained by Australians in 'priority sectors', which 'include good governance and community development training' (Australian Agency for International Development 1998). It can be extremely difficult to imagine or implement environmental, or any other form of change under repressive regimes. But a certain style of political system—one that sits comfortably alongside the promotion of economic growth and radical libertarianism—is also being promoted as part of the Agenda 21 package: American-style pluralism.

Many 'well-meaning' works prior to the UNCED 'summit' in Rio in 1992 argued that the South (and Eastern Europe, for that matter) had to become more like the USA in its political systems, if environmental degradation was to be brought under control. Underlying the development of this 'civil society' is an unquestioned acceptance of the apolitical North American form of democracy based on capitalism and global free markets. An excellent example of this mode of reasoning is found in Ghai and Vivian's (1992, pp 18–19) book *Grassroots Environmental Action: People's Participation in Sustainable Development*. They write:

The existence of a democratic space allowing the expression and defence of community rights and claims has proven to be a crucial factor influencing successful grass-roots environmental action ... The essence of these activities is to persuade or pressure the state to intervene on behalf of the communities through adoption of new legislation ...

Another key element of this form of pluralism is its aggressive individualism. As the state is deemed intrinsically apolitical, so too is the citizen. Indeed, it is only when something goes wrong that individuals coalesce into political groupings. The natural order is one in which individuals rationally pursue their own atomised needs: ultimate salvation must be sought from within, not from political action. Obviously this apolitical mentality of the pluralist system manifests itself in Australia within a whole range of Californian-style New Age environmental remedies, where the spiritual self is recentred and reinvestigated, and where the profound discrepancies between citizens' access to political power are overlooked and denied. While the New Age ecophilosophies and crystals are deemed inappropriate in the South, sustainable development still insists on promoting change to the behaviour of the people.

Here, behaviour modification is discussed within the discourse of growth logic advocated under these radical libertarian, free-market systems. Now there is a distinction between qualities of growth, not just quantity. For example, there is strong emphasis on 'capacity building', which is defined as follows:

> Capacity 21 is a novel and catalytic initiative that assists developing countries to build their capacity to integrate the principles of Agenda 21 into national development ... The ability of a country to follow a path of sustainable development is determined by the capabilities of its peoples and institutions. Capacity building is the sum of efforts needed to nurture, enhance and utilize the skills of people and institutions to progress towards sustainable development. (UNDP 1996)

This is the final insult. Sustainable development now insists that it is poor people themselves who must become more efficient. Poor training and lack of 'human development' have produced the inequities, not maldistribution or exploitation by transnational corporations and elites existing both in the North and the South. Environmental degradation can be overcome, in part, by training people globally to think in sustainable development terms. People are thus commodified and cease to have intrinsic value. This has to be one of the greatest advertising visions ever witnessed on the planet, at least in secular forms. This is Northern imperialism, using the language of ecology as its vessel. All people must perceive nature in the same way, for they do not know how to live, they do not know how to act (whether as individuals or in concert), they do not know how to think: the North must help them think and act in appropriate ways.

The United Nations Environment Program, to which the author has been a contributor, is part of this capacity building. Basically, it is a form of aid whereby most of the money is returned to the donating country in the form of contracts. The notion of power is not dealt with, there is no recognition of maldistribution of resources. Globally, the market is freed, but when it comes down to distribution of resources, the dominant agenda is once again the gross national product of the more wealthy nation-states.

And once this capacity building has 'developed' atomised humans, there can be more growth. Randy Stringer (1996), a free-market environmental economist, advocates two approaches. In fertile areas, more production must be achieved; in fragile environments, economic activities must be 'diversified'. Again, there is no notion of 'limits to growth' in certain areas. Even in fragile environments that may be unsuitable for the pursuit of economic growth, people are urged to diversify: no natural limitations are recognised. The poor, Stringer outrageously contends, 'lack the capital to invest in environmental protection'. Underlying this is the assumption that investment is innately good for the environment: once again it is sustainable development double-talk, appealing to all and renouncing nothing.

Chatterjee and Finger (1994, p 171) describe as follows the connection between business interests and sustainable development:

> Business and industry, in particular big business, were systematically built up through the UNCED process as the agents holding the key to solving the global ecological crisis. Since North and South came to agree that accelerated economic growth was the solution, [transnational corporations] had no trouble presenting themselves as the agents which could further stimulate growth, provided, however, that environmentally based trade restrictions would not impede them. Under the influence of some new management philosophies and helped by public relations, big business proposed the only intellectual novelty of the official UNCED process, i.e. 'clean' growth, 'clean' meaning technological and organizational efficiency. Not only did this eco-efficiency approach become widely accepted, but big business managed, thanks to its privileged access and its generous financial contributions, never even to be mentioned in the UNCED documents as being a problem for the environment, locally and globally.

SUSTAINABLE DEVELOPMENT, CORPORATISM AND THE STATE

At the end of the 1980s, there was still a trickle of campaigns run on traditional 'outsider' strategic lines. Antagonistic and adversarial clashes continued to flare between pro-environment and pro-development interests over, for example, the Wesley Vale Pulp Mill in Tasmania and mining at Coronation Hill (Papadakis 1993). The longer the Hawke Labor government continued in office, however, fewer of these flare-ups were ignited. Environmental concern became part of the furniture: most particularly, part of the 'round table' set up by the state and big business.

Hawke's Accord-based, consensus politics fitted in beautifully with the new resource management ideology of sustainable use. The Resource Assessment Commission (RAC), concepts of 'resource security', and the local version of sustainable development, ecologically sustainable development (ESD), emerged at this time. In 1989, Prime Minister Hawke publicly acknowledged 'the need to consider the environmental implications of development decisions through the notion of sustainable development' (Downes 1996, p 176). Some who defended Hawke argued that to include the word 'ecological' before 'sustainable development' was a major deviation from straight sustainable development; while critics felt that it was additional window-dressing. Certainly, as the ESD policy process unfolded in the early 1990s, arguments for the latter camp increased in number.

In important instances, the Australian government no longer treated parts of the movement as being outside decision-making processes. Instead, it included the movement in its policy formulation processes, convinced 'of the need to incorporate environmentalist participation at the outset' (Economou 1993, p 405). Political scientists refer to this model of politics as 'corporatism'. Like pluralism, there are many nuances in interpretations of the model.

According to Schmitter (1974), forms of corporatism have emerged in advanced capitalist economic and political systems, where pluralism has begun to decay. This decay is due to changes in the 'institutions of capitalism', including the concentration of ownership and competition between national economies. These forces have pressured the state 'to secure the conditions for capital accumulation' by intervening 'more directly and to bargain with political associations' (Ham and Hill 1993, p 40). We review later this desire to secure capital by securing resources. Due to these pressures, the state attempts to integrate powerful interest groups 'through a system of representation and cooperative mutual interaction at the leadership level and mobilisation and social control at the mass level' (Panitch 1980, p 173).

Most neo-corporatist theorists argue that this model of policy-making usually relates to powerful 'economic producer groups' such as business associations and trade unions (Schmitter 1974; Middlemas 1979; Panitch 1980; Ham and Hill 1993). Two Australian theorists, McEachern and Downes, argue that during the Accord years, certain environmental organisations also came to be included in the state's corporatist processes, though these groups are usually classified (although incorrectly) as 'consumption groups' and not as 'producer groups'. As Downes (1996, p 187) contends:

> Neo-corporatist theory can be extended beyond its traditional concern with functional, production-based interests to include organisations with the capacity to disrupt the production process and with the organisational maturity to negotiate in a neo-corporatist structure. Peak environmental organisations were included in the ESD process because of their potential to disrupt the successful implementation of development policies, a power comparable to business and labour's threat of strike.

Whereas classical pluralists saw prominent NGOs as wholly separate from the state, seeking indirect changes in the legislation through collective pressures from without, now the relationship was perceived to be closer, with key interest groups actually involved in the initial processes of agenda-setting and policy creation, instead of just reacting to what governments do.

This model of politics blurs the pluralist and structuralist definitions of the state we viewed earlier. Corporatist theorists also debate incessantly about what constitutes the state. Some writers such as Nordlinger (1981) conceive the state under corporatism as maintaining its autonomy and independence despite its close involvement with powerful interest groups at the agenda-setting stages of policy formulation. Others, such as Cawson and Saunders (1981), offer a disparate and multi-tiered view of the corporatist state. At the national level, for example, the state is perceived as a network of elites derived from the bureaucracy, business and trade unions. By including different powerful interest groups in the central processes of the state, to a certain extent these groups *become* the state. Ham and Hill (1993, p 45) take a middle position in this debate when they argue: 'Once we accept the idea of the state as an independent actor, we also need to recognise that it is not necessarily more unitary in its nature than the other participants in the policy process.'

Some commentators depict the corporatist model of inclusive decision-making as most beneficial for NGOs in general as well as those specifically within environment movements. At a minimum, it leads to the state's recognition of interest groups and their incorporation into the process of policy-making and implementation. At best, it includes environment movement elites as associate members of the state elite. Others, such as McEachern, perceive the process through less rosy glasses for NGOs and mass movements as an effort by the state to incorporate and neutralise opposition. In what is considered a ground-breaking piece of political analysis, McEachern (1993) traces the core elements of the corporatist processes as they interacted with environmental protest in Australia between 1981 and 1991. Two of these elements are referred to as incorporation and assimilation. Incorporation describes the activity where environmental activists and the business community are brought together in a set of 'normal' political negotiations. These negotiations establish the sense that there is a consensus position on environmental concern. In the act of negotiation, dissident sections of the environment movement and business (that is those who do not believe consensus is possible and/or desirable) 'are constructed as just that, dissidents against an emerging, shared, politically acceptable position' (McEachern 1993, p 180).

There is also the process in which the socially critical discourses of ecology and environmental concern are turned into 'legitimate, acceptable, non-threatening discussions about existing economic and resource development practices. The initial formulation of "sustainable development" was of this kind' (McEachern 1993, pp 180–1). We now turn briefly to the manner in which the policies of mainstream green NGOs were 'incorporated' into the state's own environmental policies, particularly with reference to the formulation and implementation of 'ecologically sustainable development' and resource security legislation.

THE POLICY PROCESS OF ECOLOGICALLY SUSTAINABLE DEVELOPMENT

The policy processes of the ecologically sustainable development process have been now well documented (McEachern 1993; Doyle and Kellow 1995; Downes 1996). We use it here to illustrate how this process differs from the pluralist and structuralist models described earlier. In late 1990 Hawke invited mainstream environmental NGOs to 'non-conflictively' participate in the ESD process along with representatives from government agencies, the Business Council of Australia (BCA), representatives from various trade unions, and the Australian Council for Social Services (ACOSS). Among the green organisations invited were Greenpeace, WWF, the ACF and The Wilderness Society. Nine sectoral working groups were established: agriculture, energy production, energy use, fisheries, forestry, manufacturing, mining, tourism and transport. The consultation process concluded in December 1991, with 500 recommendations included in numerous working group reports. These recommendations were considered by 37 state and federal committees (Downes 1996).

After receiving invitations to take part in the ESD process, green NGOs responded in different ways (see Figure 9.1). Having initially refused to nominate delegates for working groups, after some governmental compromises the ACF, WWF and Greenpeace decided to participate. From the beginning TWS decided not to participate, citing recent decisions by the Labor government to allow logging in national estate forests in New South Wales and sand mining in Queensland as the reasons for the organisation's stance. The ACF also considered withdrawing from the process over decisions to allow logging in Eden. It was the government's 'cave-in' to forest industries and unions over the resource security issue that tested the ACF's patience most. To grant the industry resource security really made a mockery of the process and undermined its validity as it guaranteed the industry's continued existence and thus pre-empted the round-table discussions. WWF and ACF stayed on in the process in the interim, but Greenpeace withdrew at this juncture (Doyle and Kellow 1995, pp 268–9).

Figure 9.1
Corporatism and the ecologically sustainable development process

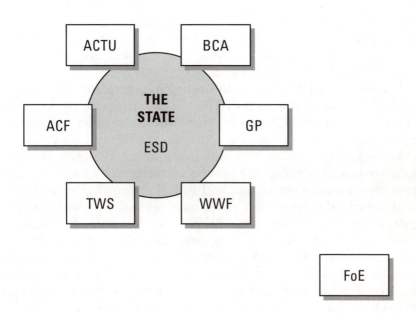

Note: Green groups responded in different ways to the Labor government's approach to ecologically sustainable development. Friends of the Earth (FoE) was not invited to participate.

Compiled by the author.

Towards the end of 1992, now with Paul Keating as Prime Minister, all environmental organisations withdrew their support from the process, claiming that the process had been hijacked by the bureaucracy. After some public assurances from Keating that he wanted the process pursued to its completion, there was a brief reconvening of the round table, with three of the four original NGOs back in the process, resulting in the ESD strategy being endorsed on paper by the Council of Australian Governments in December 1992.

In Figure 9.1 Friends of the Earth is depicted outside the process. Unlike TWS, however, FoE was not invited to participate. FoE symbolises the many organisations and informal groups in the movement that were not considered legitimate enough to enter negotiations. Selection of the ACF, TWS, Greenpeace and WWF has been justified by supporters of the ESD process on the grounds that these organisations were large and representative. Greenpeace and WWF may be large in membership, but they had few operational branches in Australia. Neither organisation had a representative structure in place. WWF was simply selected as it is the most institutionalised and politically pragmatic of all green NGOs in Australia. As we saw in chapter 5, it often mimics the industry line, providing business with green credibility in exchange for money and access.

The ACF and TWS were both Australian organisations with large memberships, albeit concentrated on the eastern seaboard. The conservation councils, which are the most representative NGOs, did not participate. One reason for this is that both TWS and the ACF had worked extremely closely with Labor during the 1980s. In fact, the ACF's patience with the ESD process certainly had a lot to do with the overly close ties between it and the ALP forged over the previous decade (see chapter 10). There were few gains for the environment or the movement during this process. The ACF, for example, had spent $350,000 of the Labor government's money to produce a report that was never implemented. The biggest loss was the exclusion of those parts of the movement that considered the cost of cooption too high a price to pay. Thus, many more radical voices had no say.

Round-table processes can be extremely useful in learning arguments advocated by opposition groups, understanding the parameters of their concerns. This can inform future strategies. So often it is useful to engage for initial periods, and then to disengage at some later juncture. Getting out before the bidding starts or staying until the shouting ends is not necessarily a signal of success. Participation in corporatist round tables ceases to be a useful strategy when it demands a conformity that may limit the range of possible responses.

The federal Labor government managed to accommodate

environmentalism into its everyday practices, and to deflate potential environmental conflict, without actually doing anything substantial in the arena of environmental protection. Resource security, for example, was portrayed as a battle between Resources Minister Griffith and Environment Minister Kelly. Griffith argued that native forests could not be 'locked up' to the point where logging interests were disadvantaged. Mill owners needed security of tenure in the forests if their operations were to remain viable. Basically, this meant that the forests were fair game. Kelly sided, in part, with major environmental NGOs such as the ACF, arguing that resource tenure had to be limited, and that the interests of native forests were paramount. As we have seen, corporatist models are often most effective as a way of securing the conditions for capital accumulation.

This was a similar scenario to the native forests/wood chip debate of late 1994 and early 1995. In this case the 'polarised' cabinet members were Environment Minister Faulkner and Resources Minister Beddall. TWS threw its weight behind Faulkner, and logging interests supported Beddall. The cabinet protagonists were different people, but the process was very similar to the resource security issue. Instead of the outsider politics of the first period, where environmentalists saw government (in all its forms) as one of their major antagonists, now— in this period of inclusion and accommodation—major forest issues were fought in cabinet, with environmental 'goodies and baddies' depicted as doing battle within the limited intra-governmental stage of Canberra.

The consequence of this deliberate attempt by the Labor Party to control environmental politics is that opposition was further gutted. Green NGOs received more access to the state through corporatist models of environmental policy-making: indeed, temporarily they became part of the state. But corporatism is more than just a theory about how the state includes selected, powerful interest groups. When environmental concerns were incorporated into politics-as-usual, they were also incorporated into business-as-usual. Ecologically sustainable development, both as an ideology and as a political process was enthusiastically embraced and sold by the Australian business community.

BUSINESS IN AUSTRALIA: SELLING THE IDEOLOGY OF SUSTAINABLE USE

There were some good wins for environmentalists during the first period, although they were piecemeal and short-lived. Business interests, in their outright rejection of environmental arguments, suffered some serious losses. Hugh Morgan (quoted in McEachern 1991, pp 15–16), CEO of Western Mining, said this:

We are barely holding our ground in this ideological war. Although we are not retreating, yet we are not winning. We need new strategies to reverse the tide that is currently moving strongly against us.

Through the 1980s environmental antagonists looked to other less conflictive means of securing their future power. No longer did many business interests across the globe deny the existence of environmental damage due, in part, to their own malpractices. Their ploy changed: to beat the bastards at their own game (but on newly defined terms and agendas), to subvert them, to divide them, to supplant them, to appear to be greener than the green. An entire raft of strategies was used. Business interests employed people to serve on 'think-tanks', which produced countervailing messages and deployed arguments that emphasised 'internal' weaknesses within 'environmentalisms' and environment movements. In addition, many green arguments were reframed largely by business within a discourse that revolved around the key concept of sustainable development. Stephen Bell (1995, p 32) discusses the details of business's agenda-setting practices in the ESD process as follows:

> This explains the role of key business associations and the work they have done as business 'think tanks' in recent years. Both the BCA and the Australian Mining Industry Council (AMIC) were responsible for promoting the concept of ESD in Australia, both in the business community and to government. The main business response has been to promote the idea that ESD can be achieved and that this will mean reliance on technology to square the circle between environment and development. The business message also stresses that 'heavy-handed' government regulation or intervention is not warranted and that market mechanisms should be used as the key policy instruments in environmental management. Thus polluting behaviour and access to the use of resources should be 'internalised' within market and price transactions in order to arrive at 'optimal' outcomes.

Some large companies genuinely attempted to alter their production techniques in the interests of environmental safeguards. McEachern provides a taxonomy of three different types of corporate responses to the environment movement (McEachern, 1991, pp 110–14). At one end of the spectrum are the 'rejectionists', who clearly 'see the environmentalists as an enemy that has to be defeated in long and hard conflict'. This is Hugh Morgan's approach, and it was the dominant business response until the mid-to-late 1980s. At the other end of the spectrum are the 'environmentalists'. McEachern (1991, p 113) describes these as follows:

> In the realm of business there are some who could be described as environmentalists or working on products and services seeking to reduce

harm to the environment. Companies developing anti-pollution devices fall into this category as do those seeking to promote genuinely 'environmentally friendly' products. There are not a great many examples.

Between these two are the 'accommodationists'. This kind of response was now the most popular among the corporate sector in Australia advocating sustainable use. McEachern writes:

> Those that have sought to accommodate themselves to rising environmental concern are many and varied. Here are businesses that see the possibility of making money by identifying green concerns, preserving their broad operations and marketing their products accordingly.

This accommodationist stance 'may well be a tactic to manage or minimise the impact of increased environmental concern' (1991, pp 112–13). It is this approach that interests us here. It is a more insidious form of attack on Australian environmentalism. These companies, now employing 'environment officers', continued to attempt to maximise profits, interpreting environmental concerns as 'a fly in the ointment' (Bell, 1995, p 30), a hurdle that can be easily skipped with the aid of some good, old-fashioned public relations.

While much backslapping occurs, and all the environmental awards are handed over to companies for, among other things, successfully trading in units made up of noxious emissions, for 'reducing' wastes from previously intolerable levels to less intolerable levels, or for sponsoring koalas (which many ecologists seriously consider are not under threat), much of the industrialised world and of the developing world experiences severe environmental degradation. Much of this degradation is produced by large corporations.

Under sustainable development, the environment is accommodated within the existing practices of industrialism. One extremely effective way has been to redefine environmental terms and to provide new corporate frameworks for their understanding. Nicole Mazur's (1995) research in relation to corporate sponsorship in zoological parks is interesting in this context. One example is the 'McDonald's Orangutan Rainforest' exhibit at the Taronga Zoo in Sydney. By being associated with both orang-utans and rainforests, the company is implicitly stating its affinity with these two 'natural objects'. The value for the company is that it is framed in the same 'picture' as these two cuddly wilderness images and hopes that customers will identify some of these 'natural' characteristics with its own product. Worse than this, the orang-utan rainforest becomes a McDonald's product. These corporations' public relations campaigns are designed to deploy 'the rhetoric of nature' (McEachern 1995) in a bid to quieten opposition and to continue to make profits.

One reason corporate sponsorship of, for example, ex situ wildlife conservation (in the case of the Taronga Zoo) is so attractive is that government funding for such projects has been reduced. Often environmental initiatives would fail if corporate funding, with all its associated trade-offs, were not secured. Business interests, although far from unified, act best in this social investment void.

One excellent illustration comes from Adelaide. In 1994 the indigenous people of the Adelaide Plains—the Kaurna—had recently acquired a piece of land leased from the disastrous Multi-Function Polis (MFP). Aboriginal people lived and thrived in what are now non-Aboriginal cities, and not just 'in the outback somewhere'. Due to the fact that the Kaurna were so completely dispossessed and displaced, it is almost impossible for them to gain anything of value from the federal native title legislation. The Kaurna Heritage Committee therefore attempted to gain funding from all possible sources to develop their site. This small piece of land has ancestral connections—despite the white ownership titles— and sits between three major arterial roads to the south of Adelaide (Laffers Triangle). These roads provide retail outlets for a number of oil companies and fast food chains. Now two of the latter sponsor the Kaurna Heritage Committee. In return, these multinational companies win community support, as well as identifying themselves symbolically as 'friends of indigenous peoples' and 'environmentally sound'.

But when resources are being competed for, this rather warm, fuzzy relationship disappears from view. Companies such as Western Mining Corporation (WMC) have a long history of direct conflict with indigenous communities both within and outside Australia. On most occasions Western Mining does not even recognise the right of indigenous communities to express an opinion over future use of their land (Rosewarne 1995) (see chapter 12).

With the 'reframing' of environmentalism by business and government interests, certain types of environmental reform began to assume prominence on environmental policy agendas. A rather narrow band of issues relating to decreased waste in production (and increased profits) came to the fore. Recycling and reduction in emissions are examples, supposedly telling us that governments were now being successful in 'greening' society. But not all environmental issues are about efficiency maximisation or increased profit margins in the medium and long terms. Earlier, very few environmentalists in Australia had been remotely interested in this type of environmentalism (to their partial detriment). As we have noted, the Australian experience had been dominated by wilderness and anti-nuclear concerns. So apart from coopting and bastardising the language of environmentalists, a coalition of business and state interests redefined the dominant agenda of environmentalism.

Some academic and public commentators imagine that the ESD process provided the only true example of corporatist politics played by the environment movement during this second period. This is not so and, unfortunately, as a case study, the ESD process is greatly overemphasised. It is true that ESD constitutes the most institutionalised and most easily identified example of Accord-style environmental politics. The corporatisation of environmental protest had enormous impacts on the everyday operation of the movement, both inside and outside formal policy-making forums. In chapter 10 we explore the ramifications of corporatist-style politics on the less formal networks of the movement involved in the latter days of the wet tropical forests campaign.

NOTES

1   This discussion of the ideology of sustainable development and Agenda 21 is based on Doyle 1998a.
2   Agenda 21 emerged from the Rio Earth Summit of 1992. It became a 'green point' for many environmental activists at institutional and corporate levels.

10

INSIDER POLITICS: THE EMERGENCE OF ELITE NETWORKS IN THE WET TROPICS CAMPAIGN

Both the wilderness and anti-nuclear movements experienced the politics of corporatism during the second period. The anti-nuclear movement was almost decimated in this period. While an ambassador for disarmament was appointed and a 'Nuclear Free Zones Treaty' was signed in 1986, the Hawke government actively supported US bases and uranium mining in Australia. Jim Green (1998, p 14) writes of this low point in the anti-nuclear movement:

> These projects also gave the leaders of the anti-nuclear groups cover as the movement declined: they appeared to be doing something. But the consequence was that the movement declined even faster as resources were diverted from important battles to reverse the government's pro-nuclear policies into providing window-dressing for them ... The ALP's coopting of the movement was facilitated by the movement's growing disconnection from the working class and the trade unions, and the replacement of a perspective of building a broad, active mass movement by liberal approaches to campaigning.

A more positive reading of the situation is found in the work of Ian Ward (1989, p 231), who argues that the uranium movement had been successful by incorporating their policies into the Labor Party platform, which resulted in the 'slowing down' of the Australian industry under the three mines position. Obviously, political incorporation can work both ways, though it usually produces outcomes that favour the powerful.

The anti-nuclear movement would not reappear as a viable force in Australian politics until well into the 1990s. In this chapter we focus on the campaign to conserve the wet tropical forests of north Queensland, and in particular on the movement's strategies in the 1987 federal election.

In 1983 and 1984 the Cape Tribulation blockaders dominated proceedings. Before then, the main arguments for conservation of the area revolved around aesthetics, wilderness, indigenousness and survival. The key networks included local inhabitants of the area, although many other environmentalists travelled from afar to attend the blockades, particularly the second one of early 1984.

In 1985 and the first part of 1986, Brisbane was the epicentre of a more formal wet tropics environmental campaign. The Rainforest Conservation Society of Queensland, the Wildlife Preservation Society of Queensland and the Queensland Conservation Council were key organisations during this period. The Australian Conservation Foundation and The Wilderness Society—two national environmental organisations—recognised that Brisbane was the campaign hub and promptly set up branch offices there. Because the campaign was now based in a northern state capital, the focus of politics was also at the state level. The movement argued interminably with the Queensland government to have the wet tropics listed as a 'World Heritage' area. Environmental arguments were now based on science and economics.

During the 18 months of the campaign before the 1987 federal election the focus moved from Queensland to the south-eastern cities. An informal elite network of up to a dozen professional environmental activists now dominated all wet tropics environmental initiatives. As a direct consequence of this dramatic shift in the movement's power base, the environment movement became totally immersed in the federal election. This was quite remarkable given that, in the past, party endorsements by environmental groups had seldom occurred. Gone were the arguments about aesthetics and spirituality, gone even were those of science and economics: the era of political and bureaucratic expediency had arrived. The movement's affiliation with the Labor government did not occur with the consent of the movement as a whole, but rather reflected the will of this elite network of powerful, professional activists.

The Australian environment movement's involvement in the 1987 Australian federal election was a 'success' in the terms of the political game as defined by the dominant, mainstream agenda.[1] Parts of the movement managed to prove their political clout in the electoral game. The swing in the voting in the 11 seats targeted by the movement was 0.89 per cent to the ALP, compared with a 1.31 per cent swing away from the ALP on an Australia-wide basis. The movement claimed, therefore, that its campaign contributed on average 2.20 per cent of the ALP vote in this election. Even if this figure is somewhat exaggerated, the most hard-headed political analysts did not deny the movement's electoral might. The movement's success can be partly attributed to the ALP's willingness to accommodate the movement's wishes in a trade-off for electoral support. More importantly, various changes to the movement's structure, goal-seeking and power distribution also played a role.

The environment movement involved with the wet tropical forests of north Queensland had become 'elitist' in its operation in 1986. Power had fallen into in the hands of the few, pressures existed within the movement to 'toe the elite's line', and the movement's agenda was determined by these powerful professionals. This situation may not appear remarkable if we were talking about a business corporation or a government body. The environment movement in Australia (and indeed in the earlier phases of the wet tropics campaign), had valued consensus and demanded that the processes of democratic representation be fulfilled before any initiative could be implemented.

Elite networks do not have to be conspiracies. Women members of The Wilderness Society Support/Review Group (TWSSRG 1988) wrote:

> They don't sit back and plot their advances. Often they are nothing more than a group of friends who are interested in/participating in the same political activities. These friendship groups function as networks of communication outside any regular channels. If there are no formal channels these are the only forms of communication. Two or more of the networks may then compete with each other. What quite often happens is that some people who are not 'in' simply opt out. They are excluded by this process.

WHO ARE THESE ELITE ACTIVISTS?

Members of this elite network dominating environmental initiatives had three essential characteristics. First, they were professional activists. They were not volunteers, nor were the majority honorary elected officials: they were employed to do a job. Second, they worked for either the ACF or TWS. Some of them had ties with both formal organisations. Third, all members were located in one of the four major capital cities: Canberra, Sydney, Melbourne or Hobart (see Table 10.1). The first three of these cities reflected the 'triangle of power' that is the nexus of mainstream politics in this country. This elite network of environmentalists therefore mirrored the establishment's distribution of power. The addition of Hobart reflects the history of The Wilderness Society, originally the Tasmanian Wilderness Society. Despite the intense involvement of the Queensland groups in previous years, and the fact that the wet tropical forests are in Queensland, not one Queenslander was included in this network.

TABLE 10.1
THE NETWORK OF PROFESSIONAL ELITE ACTIVISTS

Name	Organisation	Position	City
Jonathon West	TWS	Director TWS	Hobart
Karen Alexander	ACF, TWS	Councillor ACF and election officer TWS	Melbourne
Michael Rae	TWS	Coordinator TWS	Melbourne
Phillip Toyne	ACF	Director	Melbourne
Bill Hare	ACF	Campaign director	Melbourne
Jane Elix	ACF	Campaign officer NSW	Sydney
Margaret Robertson	TWS	Coordinator TWS	Sydney
Judy Lambert	TWS	Liaison officer	Canberra
Joan Staples	ACF	Liaison officer	Canberra

Note: I arrived at this de facto network by asking several key individuals in the movement to nominate the network participants who dominated environmental proceedings in the election campaign. By cross-referencing these names, I compiled this list of the most powerful individuals. (Of course, each person I talked to nominated a slightly different list, reflecting their different individual locations, biases and relationships.)

These professional environmentalists, and others like them, devote their lives to environmental causes. They accept wages far below what they could earn elsewhere given their skills. They perform in a way they believe benefits the movement as a whole. Their dominance of the movement in this instance was not necessarily the result of a conscious bid for power but a consequence of their attempts to play the political game as defined for them by party-political and government agendas. It was this network of individuals who bargained with the Labor government before and during the 1987 election campaign. The decision was theirs. Due to the key positions of power this national elite held, they were able to convince the politicians, the media and the general public that their actions reflected the wishes of the movement in general. As we shall see, this was not the case.

UNREPRESENTATIVENESS

The decision to play the electoral game was first made by TWS and later endorsed by the ACF. In the past, the ACF had been the front runner in electoral involvement. In the 1987 election, however, TWS had the upper hand all the way. TWS's decision to back Labor in the House of Representatives and the Australian Democrats in the Senate was announced publicly on 6 June. That decision had already been made by the TWS elite band sometime before the national meeting of TWS held in Brisbane on 28–29 March 1987.

This meeting constituted an attempt by the organisational elite to gain ratification from the TWS membership for its electoral strategies. The voluntary membership was quite shocked at the extent of preparation that the organisation's professionals had undertaken. Two extensive documents had been prepared by elite network members Michael Rae (TWS Melbourne) and Geoff Lambert (TWS Sydney) for discussion papers prior to the national meeting. In short, the elite had set the agenda. One TWS volunteer wrote of the decision at the national meeting to play party politics as follows:

> All those supporters of TWS, like myself, who believe that TWS should take no party political stance at all could very rightly feel angry, misrepresented and wonder just whose decision it was that they were abiding by and, by default, condoning ... that really wasn't consensus decision-making at all. I see TWS as losing what has always been its most attractive quality, that of grass roots involvement and decision-making. Power it seems is vested in those who are paid by TWS rather than the members (Morgan-Thomas 1987).

Indeed, few volunteers were admitted to the network, but this is not the full extent of the division between the elite group and others. The organisation's professionals operating in Brisbane, Cairns, Adelaide and Perth were equally surprised. Harry Abrahams (1987), coordinator of the Brisbane branch of TWS wrote that: 'The Brisbane Branch was not thinking strongly about elections before the meeting, hence some members may have felt there was a bit of "steamrolling" going on.'

This strategy of not consulting organisational workers 'outside' the network continued throughout the campaign. The Brisbane employees of both the ACF and TWS presented a brief report for the campaign post-mortem held in Melbourne the week after the election. They voiced their disappointment about being left out of the process of power-broking and decision-making:

> We were not told who was working where and what numbers were to be used to contact whom when the branch offices were established. We found out by hook or by crook rather than being told. Somehow we felt that we were not aware of where and when decisions were being made in Melbourne. Did regular meetings exist between the ACF and TWS? We felt more that we were finding out the decisions after the meeting rather than that a meeting was going to be discussing ... this, what do you think? (Abrahams, Bond and Wilson 1987).

So, at some stage before the national meeting of TWS, the professional elite activists had made two decisions: i) to play the electoral game, and ii) to back the Labor Party. Proof of the second point is easily found in Geoff Lambert's (1987) report to the national meeting. He writes:

> This morning's [23 March 1987] report that Howard (leader of the Liberal Party) would abolish everything including motherhood, if elected to office, opens up greater scope for differentiating between Liberal and Labor in the eyes of the conservation-minded voter. Perhaps Labor is a viable choice. Howard has turned rather dry on conservation, as evidenced by his promise to abolish support for the ACF (and us too?).

But the decision to back Labor goes even further back. Until November 1986, Jonathon West worked for Barry Cohen—then Minister for Arts, Heritage and the Environment—as his private secretary (Ruzicka 1987). West's timely move to the directorate of an organisation that would take the lead in the environment movement's election campaign was far from a coincidence. According to one Tasmanian TWS source, the decision to appoint West director of TWS had been made some 18 months earlier by the TWS elite. It was just a matter of timing.

West became the self-appointed leader of the movement during the election campaign. Trade-offs and deals were achieved through his negotiations with both parties. It is irrelevant to ask whether the movement or the ALP was more important to Jonathon West. Certainly he played a key role in the movement's election campaign and the consequent endorsement of the Labor Party. Although the environment movement cannot be delineated on party-political grounds, this elite network—particularly the TWS members—was firmly ALP-oriented. The Wilderness Society's move to focus on the 1987 federal election, and the consequent support it gave to the Labor Party, did not reflect the express wish of its membership or of the movement as a whole. TWS did not operate under a tightly constructed constitution during this time. The constitution in force in 1987 was not designed to restrict powers. Section 7, paragraph VIII of the constitution read: 'Any four members of the committee shall constitute a quorum for the transaction of the business of the committee.'

Four people can 'exercise all such powers and functions that are required by the Association'. The inbuilt flexibility of TWS's constitution had been used in the past to act quickly in certain crisis situations. Unfortunately, without these constitutional controls its organisation had been seized by a small network that did not seek to be representative; instead, its members believed that they 'knew what was best' for the rest of the society.

In the past, it had been the ACF that had dictated the national environmental agenda. The ACF voiced its support of the Labor Party officially on 15 June. The Adelaide *Advertiser* reported the next day that:

> The 20,000 strong Australian Conservation Foundation will endorse the Hawke Government and the Australian Democrats in the election campaign ... The decision, based on a two-thirds majority vote in the ACF secret ballot of its members, is a big plus for Labor, which has made several pledges to woo the conservation vote ...

The general public, on reading this story, would have been convinced, wrongly, that two-thirds of ACF's 20,000 members had voted to support Labor in the election. In both previous elections, the ACF membership was polled to ascertain whether or not the ACF should back a political party. In the case of the 1987 election, however, the ACF did not adopt this policy. Instead, the broad decision to back the Labor Party was left up to the ACF council, while the nuts and bolts of the campaign were formulated by the seven-member executive. The ACF argued that this latter process saved time and money. It also substantially disempowered its membership.

The response of the West Australian chapter of the ACF illustrates both the lack of representativeness demonstrated by the council's vote and reaffirms the point that the elite power nexus was based in the four south-eastern capital cities. Colin Hall (1987), president of the chapter, stated that:

> We were notified of Council's decision three weeks before the election. We weren't consulted. We are sick and tired of not being consulted. We never see the ACF Councillors at any of the meetings. Some individuals within the chapter challenged the Council's decision. But they were heavied out of this stance by Labor Party sympathisers. The long term effects of the electoral campaign are now being felt, the whole affair has split the branch into disrepair. I believe it will fold.

The ACF council's stance merely ratified the elite network's election strategy. One ACF councillor, who did not want to be named, made the following statement about his role in the Labor Party endorsement: 'We had been lobbied heavily before we cast our vote. It was a secret ballot, but in reality it was a foregone conclusion' (personal communication to author, 1987).

The ACF members of the elite network, unlike their TWS counterparts, did seek some sort of ratification from the broader membership. But this was merely a formality. The decision to combine the electoral efforts of both organisations was made at a national meeting of elites in May 1987. Eight members of this network decided 'that the organisations should cooperate as far as possible in formulating a common platform for the election in relation to the question of support or otherwise for any political party' (Hare 1987).

The opinion of the former director of the ACF, Geoff Mosley, is most interesting in this context. Mosley saw the changes in the ACF's election procedures as just one example among many of the increased elitism in the organisation and 'the continuing, progressive reduction of the powers of its general membership' (Mosley 1987). This lack of consultation was rarely a deliberate ploy of the professional elite. Some of the individuals involved were equally concerned about this increasingly unrepresentative decision-making process. The network was dictated to by the short-term time frame and the agenda of the political parties. By playing the electoral game, the structure of decision-making and the scope of the movement's goals were dramatically affected.

THE ALP CONNECTION

The ALP had a substantially better environmental policy than the Liberal or National parties. The 'necessary and sufficient conditions' for movement support of the ALP were spelt out at the Sydney meeting of the elite group in late May:

1   Unilateral nomination of the wet tropical forests and the commitment to use Commonwealth powers to stop the degradation of this area (logging, roads, real estate, etc.)
2   Injunction to stop illegal logging and forestry operations in Tasmania. (Hare 1987)

The ALP promised to meet these two criteria, thus satisfying the movement elite that a substantial bargain had been struck.

This both simplifies and clouds the reasons for the elite group's endorsement of the ALP. If policies and past performance were the only criteria, then the Australian Democrats should have received full support. The issue of ALP endorsement by the movement elite was far more deeply entrenched. There is no evidence to suggest that the ALP deliberately infiltrated the movement. There was no evidence of conspiracy. Yet from 1985 to 1987 it attained increased access to movement politics.[2] This is explained by a number of factors. First, we might recall the role of Jonathon West, who played 'the go-between'. West was directly influencing the movement's path, even while he was still in Barry Cohen's employ. Working for Cohen, West used ACF and TWS letterhead when expressing his views to the organisational elite. One such paper, entitled 'The Wet Tropics: What can we expect from the Labor Government in its second term', is an excellent example of his dual role. This paper, written at the beginning of 1985, begins with this brief paragraph: 'This paper aims to set out briefly some thoughts on how the re-elected Labor Government might react to the tropical rainforest issue in 1985 and how conservationists can orient in the new circumstances' (West 1985).

Apart from supplying advice to movement participants, West directly involved himself in the organisational elite power-plays of the environment movement. The dismissal of Geoff Mosley in April 1986 has been popularly portrayed as a power-play within the movement between 'radicals' and 'conservatives', Mosley belonging to the latter. This line of thought proceeds along the following lines: after 20 years of service to the ACF, Mosley was overthrown in a bloodless coup, very reminiscent of the situation that saw the wholesale changes in the foundation—and his rise to power—in 1973. The reality is quite different. Mosley was removed from his position by the same elite network that now had a firm grasp over the two national organisations and the movement as a whole. Quite simply, Mosley was a sometimes abrasive and often idealistic character who would not kowtow to politicians and bureaucrats. The president of the ACF, Hal Wooten (1986), expressed to ACF council his concern that Mosley was not creating sound relationships with government officials: '... the growing alienation of those with whom ACF must maintain relations—ministers, public servants, and the leaders of other conservation organisations'.

The professional elite had put Mosley's dismissal on the agenda and council merely ratified its decision. In the words of long-time ACF councillor John Sinclair, council had become little more than a 'rubber stamp', 'that means that the major decisions are predetermined outside the Council'.

Mosley did not get on with the minister, Barry Cohen. This situation was seen by the elite as being totally unsatisfactory. One ACF employee summed up the dismissal of Mosley quite simply: 'Jonathon West was on the phone for the week before the Council made its decision.'

West had operated in the federal party-political forum, and his outlook remained based on this mainstream 'appeal to elites' agenda (Martin 1984). He brought the ALP much closer to the movement. In the words of Michael Rae (1987), (TWS convenor, Melbourne), West 'has provided us with far more access to the Labor Party political machine and the higher echelons of the bureaucracy'. It is interesting to note that West resigned from the directorship of TWS immediately after the election, giving very little notice to the organisation. The *Times on Sunday* reported: 'Mr West did not stay long on the job as Director of the Wilderness Society after assisting the Hawke victory in July. A few weeks ago, he resigned to become the Canberra economics and foreign affairs writer for Rupert Murdoch's Melbourne *Herald* (Higgins 1987).

The relationship between the professional elite and the ALP during the election was far closer than just a mutual back-scratching exercise. Some members of the elite network were also ALP members. But this was not the crucial factor in the link between the movement and the ALP. Highly ranked party officials became part of the movement's elite network during the election campaign. Bob McMullin (federal secretary of the ALP), Peter Batchelor (secretary of the Victorian ALP) and Peter Beattie (secretary of the Queensland ALP) were some of the key Labor members of this network, and were involved in discussions relating to movement strategies.

As evidence of this direct involvement of the ALP in the movement's affairs, we need look no further than the issue of seat selection. The initial selection of seven key seats that would be 'king hit' by the movement during the election was not a movement decision. The Wilderness Society election evaluation report spells this out quite clearly: 'First decisions (about seat selection) were made at ACF/ TWS "Sydney Meeting" on June 6 on advice from ALP Secretariat' (Alexander 1987). Some professional members of the movement 'on the outside' were not at all amused by this procedure. Abrahams (1987) writes: 'Our move into Fisher [an electorate] was done very quickly and we were upset that the decision was made without consultation with those of us closest to the scene, not good. We were happy with the decision at the time but not the process.'

And not only did the movement endorse the ALP during the election, but their own campaign focuses were initially defined for them by the ALP and, in certain electorates, the movement's campaign became almost inseparable from that of the ALP. The electorate of Denison in Tasmania is a classic example. Nowhere else was the ALP–movement link stronger than in Tasmania. The swing of 4.69 per cent away from the sitting member Michael Hodgman was more than enough to put his ALP opponent, Duncan Kerr, into the House of Representatives. In this electorate, TWS did a substantial amount of the ALP letter-boxing (Holloway 1987).

One further reason for the increased interplay between the ALP and the movement was that the ALP was in government. In the 12 months before the 1987 election, the professional elite concentrated its efforts more on direct lobbying techniques aimed at influencing powerful people in the mainstream political sphere. The days of mass mobilisation campaigns, for which TWS was renowned, were over. The power of the movement was now in the hands of a small network of professionals who were far more interested in dealing with their counterparts in government than generating grassroots action. Thus, the movement had moved closer to government during this period: the ALP just happened to be in office.

This professional elite network spoke the language, used the same arguments, and began to think in the same way as the political leaders. No more arguments about wilderness, no more talk of scientific diversity, instead the game was mainstream party politics: deals, bargaining, pragmatism and money. The movement was now playing the political game as defined for them by the dominant power-brokers. To do this properly they needed money—large amounts of it. The source of movement funding during the election campaign deserves a mention here as it illustrates the ultimate authority of the elite network, who, on their rise to power, brought with them an ideological package reflecting mainstream values. Means were not especially important to this group. Ends—short-term ones—were the priority.

CORPORATE SPONSORSHIP

For the first time in the history of the Australian environment movement funds were coming from large corporate sponsors. In the past, the movement had relied on individual donations, membership fees, government grants and its own pockets. In the 1987 federal election

Corporations are increasingly funding Green initiatives in a bid to gain governmental kudos. *Courtesy of Friends of the Earth (Australia)*

vast sums of money came from corporate sponsors in a bid to influence the outcome of the election in favour of Labor. This occurred for two major reasons. First, the largest and most powerful corporate bodies in Australia backed the ALP, perhaps one of the few times the party has enjoyed such support. Second, the movement—or those networks that dominated it—had witnessed a fundamental change in the structure of its power distribution. This transfiguration was matched by dramatic ideological changes.

For example, to this day most members of the society remain ignorant of it. In the draft election evaluation document that TWS prepared for its members, there is no reference of this donation. In fact, under the heading 'Donations', only $11,000 is recorded, coming from the generous pockets of the general public (Alexander 1987). Another $70,000—again this was unrecorded—was accepted by TWS straight out of the ALP campaign fund. This disclosure was made over a year after the election in Brian Toohey's publication the *Eye* in November 1988.

At the 1984 TWS national meeting, Pam Waud, a prominent environmentalist, presented a discussion paper demanding that certain ideological questions be resolved. At that stage TWS was caught in an ideological dilemma, illustrated by the conflict between its grassroots members and its professional bureaucrats. Waud (1984) asked of her members:

1 Do we regard TWS as an end in itself (as an organisation) or, is it a structure created by a changing group of dedicated individuals AS A CAMPAIGN TOOL?
2 What is the individual's place vis-a-vis the organisation, and the efficiency of the organisation?
3 Should we seek and/or accept funding from environmentally unsound organisations/individuals?

Waud's questions were not resolved at the 1984 meeting. They were, though, in the 1987 election campaign. The decision to accept corporate sponsorship epitomised this transition. TWS—the organisation most dominated by this network—had finally resolved the ideological contradictions that had existed since the fulfilment of its Franklin goals: the 'organisation' had won. Secrecy now existed between this elite network and the rest of the environment movement. Files were off limits. All was done on the phone. The professional elite group now controlled information flow.

ANALYSIS

During the 1987 election campaign, a professional elite generated within two formal organisations unilaterally acted as representatives of the environment movement. This changed the membership—along with the ideology—of the environment movement in Australia. The organised environment movement became more narrow in its base and less ideologically diverse. The president of the ACF, Hal Wooten (1986), supported this contention when he wrote: 'But actual party endorsement is divisive and may narrow our base. It brings all the problems of single issue politics. Do we need it?'

Also, the organised movement became far more homogenised in its belief structures, as the dominant group continued to demand more uniformity in environmental ideals. In a letter withdrawing her services from The Wilderness Society, Eleri Morgan-Thomas (1987) wrote of the lack of ideological tolerance within the organisation during this period: 'Brisbane Branch has been having some very real problems lately, not the least of them being communication ... Many of us at the National Meeting felt that it was difficult to make a comment that disagreed with what was perceived to be the "power base"...'

The demand to conform to the professional network's ideology alienated large sections of the organisational membership and the movement participants who choose to operate outside the structures of formal organisations. In turn, this alienation led to polarisation and conflict. Most alienated were the voluntary workers in the more formal groupings, those people who involve themselves in the grassroots activities of any environmental campaign. Apart from being tied to particular political parties, they lost their power: a feeling of uselessness prevailed. Craig Jones (1987), another volunteer who resigned at this time, wrote to the director of the society explaining his disappointment:

> I regret to inform you that I can no longer remain a member of The Wilderness Society. I've delayed in writing this letter for several reasons, not the least of which is the remorse that I felt at having to withdraw my support for The Wilderness Society.
>
> I believe that the 'raison d'être' of The Wilderness Society was that it provided a basis for a 'grassroots' conservation movement ... With increasing executive power, which seems to be the direction in which the Society is heading, comes a limiting of the ORDINARY MEMBER's ability to participate in any way but a superficial manner. Not all of us are involved in The Wilderness Society simply to ease our consciences. We need to be active, and we need to be able to participate in the decision making processes of the society. Without this kind of participation, the Society will wither and die.

The letters of Morgan-Thomas and Jones are not unique. After West's election as director, TWS continually alienated its volunteer members in the various branches. TWS was becoming increasingly centralised. According to Geoff Holloway, a long-time commentator on The Wilderness Society, in 1986–87 the number of active TWS branches dropped from approximately 40 to 20. This was, arguably, a direct result of the branches no longer being involved in the decision-making process.

A similar trend occurred in the ACF. Earlier we saw Colin Hall's assessment of how the Western chapter had been torn asunder. Similar events were occurring in most capital cities: the organisational membership was changing. And who was replacing the volunteers who were leaving? Karen Alexander (1987), a professional, and a member of both the ACF and TWS, reported that the ACF attracted about 300 new members during the election campaign. She asked: 'Are we just attracting ALP voters and Left-wing fanatical green anarchist guerrillas? ... We could have ...'

Her reference to an increasing influx of Labor Party supporters is a most interesting point. Could it be that the movement, through continually aligning itself with the Labor Party, was becoming increasingly attractive to Labor Party members? If so, the movement's environmental agenda was also becoming increasingly dominated by party politics.

An oligarchy arose within and between two formal environmental organisations, but not in a multitude of other collective forms that also operated in the wet tropical forests campaign. Moreover, these two organisations happened to be powerful enough to project themselves as *the movement*, not just one extremely useful part of it. When organisations come to be dominated by an elite—such as the ACF and TWS during the wet tropics campaign—they no longer respect or, perhaps, understand, the complex mechanisms that combine to create the many diverse and amorphous parts of social and political movements. Without this understanding, formal organisations and their leaders will continue to force their own brand of politics upon other equally important parts of the environment movement, thus severely limiting its political diversity and forms of 'success'.

Chris Rootes (1987) writes of this limiting focus in the context of a brief critique of Alain Touraines:

> In Touraines' tidy minded attempt to abstract a single transformative social movement from the plethora of fragmentary protests and campaigns, he omits to consider the possibility that social and political change is more often the product of the many pin pricks of uncoordinated protests than it is the deliberate achievement of a self-conscious social movement of the kind he idealises.

If this trend towards elite dominance—with its emphasis on electoral politics—were taken to its logical conclusion, then in the end the bureaucracy, the government, the political parties and corporations would have complete control over the movement's political agenda. The time dimension, the rules of the game, the extent of the trade-offs, the sources of money—all these factors would be defined by the dominant regime.

The mainstream wilderness networks of the environment movement flourished during the second period, while the anti-nuclear movement waned under the three mines policy of the ALP. Organisations such as the ACF and the WWF were finally recognised as being 'part of the family', glowing with new-found respectability, and enjoying the financial rewards that come with this. Many movement networks, however, did not survive the mid-1980s hiatus. With the widespread perception of movement 'success' (short-term, as it turned out), it was even more difficult for informal, radical networks to structurally critique the operations of the state and big business, and to dramatically challenge the ideology of unrestrained use.

This is not to say that more structuralist, feminist and other forms of radical critique completely disappeared. Networks refusing to deal with elitist politics still existed, but were often driven underground. Even within TWS there were networks that challenged the professional elite dominating the organisation. For example, in March 1988, a group of women who worked in the Hobart branch of TWS produced a restructuring paper. From this restructuring paper, as we noted earlier, a 'Support/Review Group' was formed, which put into place many of the democratic structures that had been previously overridden in the larger organisation (TWSSRG 1988).

Despite the continued existence of these more radical traditions, it is fair to say that the mainstream resurgence in the wilderness networks of the environment movement only occurred in those parts whose policy-makers were willing to swallow the ideological pill of sustainable use and development.

The umbilical cord tied between them meant both the movement and the ALP would share a similar plight. In 1987, luckily for the movement—and the last vestiges of the wet tropical forests they bargained with—the 'right' election result occurred in 1987. But what a risk: a risk taken without consultation with ACF and TWS members, let alone the thousands of other individuals, groups, organisations and networks—not affiliated with either organisation—who had fought, in their own way, to save the wet tropical forests over the past decade.

In the federal elections of 1990 and 1993, similar deals were struck between professionals in the movement and the ALP, though it must be said that under Keating the Labor–Green accord was no longer seen as important. As I argued in chapter 6, social movement wins tied to party politics produce only short-term gains. Creating such intimate ties with the ALP also meant making opponents in the conservative coalition. With the defeat of Labor in 1996, the environment movement would have to face up to the consequences of that choice.

NOTES

1   This study has been published in abbreviated and different forms elsewhere, most recently in Doyle 1991. It was some years before I managed to publish these arguments in a movement journal. Interestingly, most of the arguments against dissent within the movement (see the introduction) emerged in personal and public correspondence relating to publication of this material.

2   For a revealing analysis of the ALP–environment movement relationship during this period from the perspective of Labor see, for example, Kelly 1994.

THE THIRD PERIOD: WISE AND SEQUENTIAL USE (OR THE GANG BANG THEORY OF NATURE)

The attack on the environment movement by the conservative coalition government, in close collusion with powerful business interests, has been vicious and without precedent in the last 30 years. Since its election in 1996 the Howard government has attempted to disempower the environment movement and to discredit its concerns in a number of ways. First, it has renamed its own environmental agenda a 'brown' one, with most of the funds allocated to 'environmental' projects from the partial sale of the previously state-owned Telstra being directed towards rural, primary industries. Second, it has removed or reduced funding to its most vociferous critics such as the Friends of the Earth and the Australian Conservation Foundation, while additional funds and support have been directed to the politically palatable and more narrowly focused 'nature conservation' organisations such as International Humane Society (HSI) and World Wide Fund for Nature, and several of the more traditionalist conservation councils. Wherever possible, it has removed globally recognised green issues such as greenhouse and forests from its national agenda (Doyle 1998b).

The government has decimated two decades of environmental legislation, and has replaced it with the Environmental Protection and Biodiversity Conservation (EPBC) Act, which hands over large slabs of Commonwealth determination on environmental issues to the resource exploitation-oriented states. This Act defines environmental concern in a constrictive 'Noah's Ark' form, focusing on threatened species, the RAMSAR convention for protecting birds in wetlands, and the magic word 'biodiversity', used as a symbol of inactivity. A fundamentally apolitical agenda has been created; therefore, it does not challenge business and politics-as-usual.

It must be said, however, that Howard's savage assault was made possible by Labor's retreat from environmental issues under Keating. Particularly after the election campaign of 1993, fought largely on economic issues (and the potential impact of the Fightback! package), Keating felt the ALP no longer owed a debt to the environment movement, and he would only deal with the movement if it were 'government-friendly'. He commented to staffers after his 1993 election win: 'Now the environment is back where it belongs' (Simpson 1996c), meaning now it was a non-issue, compared with the 1987 and 1990 campaigns when the green vote proved important in the re-election of the ALP. This era contrasts with the cosy relationship the peak environment groups, and especially the ACF, had with the Hawke government under the leadership of Phillip Toyne and other elite activists within the movement.

Greenhouse research is an excellent example of this new barbaric approach. Greenhouse is accepted in most nations as a crucial shared concern. But Environment Minister Robert Hill's performance at

Kakadu Wetlands in the late 1990s. Kakadu symbolised the new wave of environmentalism in this country; one that incorporated the issues of indigenous rights, anti-nuclear and traditional wilderness concerns. CCSA

Kyoto at the Greenhouse Summit, despite being testament to his diplomatic skills, was evidence of increased government pandering to corporate interests in most areas of environmental policy development. In fact, so corrupted was this process by the pressures of vested business interests that Alan Powell, a key technical adviser to the Australian Bureau of Agricultural and Resource Economics (ABARE), resigned. ABARE had placed a price tag of $50,000 on joining the policy defining committee. This effectively promoted business input and denied entry to most NGOs. To make matters worse, these sources of funding were kept secret. The Commonwealth Ombudsman, Philippa Smith, stated that: 'By not allowing adequate and balanced community input, and by not accurately declaring the sources of its funding in its climate change report, the bureau has compromised the credibility of its work' (Smith quoted by Lunn 1998, p 2). This hijacking of the national greenhouse agenda by business interests is also evident in the Howard government's cessation of funding to Macquarie University's Climate Change Centre, which resulted in its closure.

The government continually advocates free-market, radical libertarian solutions to environmental problems, diminishing the role of active public and community sectors in environmental monitoring, regulation and problem-solving. In contrast, the input of large corporate interests into the shaping of government policy is ever-increasing. This free-market ideology, ushered in by the Labor Party, but in a far purer form under the coalition, now threatens environment movements directly.

The coalition, ably assisted by the Democrats in the Senate, is advocating US-style resource management techniques and decision-making styles fashioned and promoted by what is known as the 'wise use movement'. The days of government-centred, pluralist and corporatist-style decision-making are no more under wise use. The roles of labour unions, the Public Service, and 'independent' or 'bureaucratic' science have been minimised in these resource management equations. As the market place was deemed 'natural' under sustainable use, wise use builds on this premise, and then reverses it, arguing that nature itself is a free market.

THE IDEOLOGY OF WISE USE[1]

Apart from assailing the notion of 'The Commons', the wise use movement advocates a range of instrumental attacks on environment movements. These have been ably documented by critics of the wise use movement.[2] They include the creation of right-wing think-tanks that produce propaganda challenging dominant myths of green movements, the use of 'dirty-tricks' campaigns (designed to 'smear' the reputations of its opponents), and the formation of anti-green front groups (masquerading as environment groups to confuse the general public. I shall argue here that the wise use movement has also championed a particular style of decision-making that it believes gives it more power over decisions relating to the utilisation of nature, while it masquerades behind a sales pitch that praises the supposedly democratic nature of its decision-making.

The wise use movement emerged in the late 1980s in the USA, but did not emerge in Australia until the mid-1990s. One of the founders of the movement was Ron Arnold, who argued that business could not survive the attacks made on it by environmentalists unless it, too, took on the attributes of a social movement. This movement is largely rural-based, 'anti-environmentalist', 'localist' and populist. It shares some salient features with other right-wing movements, or lobby groups that have evolved out of America's west. For example, it shares with 'The Freemen' and the powerful gun lobby the notion that the 'gumment' (government) almost always serves the interests of liberals 'back east', and that government impinges upon individual rights, such as the right to bear arms, or the right to do anything to private property. Indeed, public ownership of forest lands, for example, is also 'anti-constitutional' according to these groups and, in this sense, the wise use movement is radical libertarian in its economic focus, and pathologically conservative in its morality.

Proof of the wise use movement's advocacy of radical libertarian economic systems is abundant. Ron Arnold, for example, has also served as executive director of the Centre for the Defence of Free Enterprise (CDFE). This think-tank, according Mark Dowie (1995, p 93), an American critic of wise use, formulates 'an ideology for a grass roots insurrection he (Arnold) believes will save free enterprise capitalism from the scourge of environmentalism'. The wise use movement is backed by large extraction corporations, including Exxon, Louisiana Pacific, and Boise Cascade. The key movers on the ground, however, are right-wing lobby groups, including the Moon-affiliated American Freedom Coalition (Rowell 1996, p 190), and thousands of 'populist, small-town, and rural citizens' groups' (Dowie 1995).

MULTIPLE USE MODELS OF RESOURCE MANAGEMENT

Ron Arnold and others coopted the term 'wise use' from Gifford Pinchet, an early twentieth-century icon of the forest service. Arnold deliberately chose it for its ambiguity and its positive connotations of 'wisdom'. Pinchet championed 'multiple-use' as a way of resolving land-use conflicts. These models were firmly rooted in notions of nature as constituting a series of resources for human use. Nature itself was not of value outside these utilitarian notions. Multiple use has been dominant as a resource allocation model in many parts of the world, including Australia. It emerged most strongly in the language used to establish the Great Barrier Reef Marine Park in the 1970s. It was during the second period, however, that it became *the* key resource management strategy within the ideology of sustainable use.

Multiple use planning imagines nature divided into a pie. All possible uses of nature are known and explored, as all of nature is a commodity. It is also construed as a series of pie segments. Multiple use is reliant on pluralist and corporatist concepts of power and the state. The round table is a symbolic representation of the pie, the table itself being provided by the state. People, or the now more fashionable 'stakeholders',[3] are physical manifestations of possible resource use and conflict. Environmental managers using this multiple use model argue that most environmental processes are reversible, and that all sectors can pursue diverse goals for resource use without unnecessary conflict. All it takes to succeed is for conflicting sectors to produce end positions based on compromise and negotiation. We will return to this later.

Multiple use decision-making has enormous weaknesses. In true pluralist terms, all interest groups are perceived as equal stakeholders, while the state portrays itself as an objective facilitator, attempting to provide a working compromise between conflicting sets of 'values'. Values, of course, are important, but by concentrating solely on reconciling values, power differentials between conflicting positions are almost totally overlooked. Free marketeers see multiple use models of decision-making as essential to promote market-based and market-advantaged outcomes. Indeed, one of the 'birthdays' of the wise use movement occurred at a three-day 'Multiple-Use Strategy Conference' in Reno, Nevada, in August 1988 organised by the Centre for Defence of Free Enterprise' (Helvarg 1994, p 76).

To be fair to multiple use resource allocation and decision-making, when genuinely interpreted as a conservation regime it can achieve results. What is essential to good multiple use decision-making is the open recognition that some interests, some values, are more important than others and that not all resource issues can be 'win-win' situations. On many occasions, certain interests are more successful than others. The example of Australia's Great Barrier Reef is an excellent one. Central to the reef's management is the existence of marine national parks, or 'no-take areas' (Prideaux et al. 1998, p 15). The benefits of no-take areas are widely recognised by scientists attempting to maintain some 'control' areas that allow the development of successful management strategies. The majority of the Barrier Reef is not so exclusively protected. Most of its Marine Protected Areas (MPAs) allow commercial interests to enter the reserve. Multiple use can only work from a conservation perspective when these multiple access areas are coupled with a strong sample of representative habitats that are excluded from commercial activities such as mining. Frameworks with no exclusions are at best ecologically useless and, at worst, promoters of commercial, anti-ecological interests. These systems cease to be multiple use and become wise use.

WISE USE, SEQUENTIAL USE, AND THE GANG BANG THEORY OF NATURE

Wise use has many similarities to sustainable and multiple use. In some ways it is simply a reinvention of multiple use, but it is larger than life, a more brutal, less subtle manifestation that deliberately distorts the earlier concept. It is this ideology of *use* that has emerged so forcefully under the Howard government.

In wise use round tables there is no labour 'stakeholder'. Under the Accord processes of the Hawke government, union interests were seen as integral to resource decision-making. Under wise use they are usually absent, now seen as a non-constituency, a non-stakeholder. The exclusion of unions shares some similarities with Thatcher's United Kingdom when, during her 11-year rule, 'trade unions were dismissed from the "triangular" relationship' between the state and business' (Ham and Hill 1993, p 440).

The state has almost no role under wise use. In multiple use models such as those used in the period of ALP-led corporatism, the state is portrayed as an arbiter at the centre of the wheel or table. But the central role of the state within pluralist and corporatist models (or even the pretence of state neutrality) is bypassed in wise use decision-making. The state is usually just another participant in round-table discussions, or, worse, the interests of large business and the state are so closely intertwined that the two are indistinguishable. The state as convenor is now dead.

Corporations are now the carpenters of the round tables, initiating and controlling agendas and terms of reference. Worse still, they are largely self-monitoring and self-regulating. One excellent example of this is the Olympic Dam Community Consultative Forum (ODCCF) set up by Western Mining Corporation, the federal government and the South Australian government to 'monitor' the expansion of the uranium and copper mine at Roxby Downs to become one of the biggest uranium mines in the world (see chapter 12).

'No-take areas', exclusion zones, scientific control zones and ecological buffer zones are now increasingly obsolete. Under wise use decision-making, these areas are now referred to as resource 'lock-ups'. Any ecological 'use' is only considered *after* business interests have determined they have no further use in mind for the areas in question. Other wise use jargon refers to this as 'sequential' use, which promotes a strict hierarchy in the order of access to the 'resources'. The biosphere, in this view, can be used over and over again, fulfilling all the demands placed on it by the multitude of stakeholders, with no long-term negative consequences. This can be better understood as the 'gang bang theory of nature'. Environmental interests and those of indigenous peoples are most often given access last.

GREAT AUSTRALIAN BIGHT MARINE PARK: AN EXAMPLE OF WISE USE PLANNING

The earth's terrestrial environments are largely perceived by current wise use resource extraction regimes as only two-dimensional: only the surface is regarded as a place where multiple uses have to be accommodated, whereas sub-surface, environmental concerns are regarded as non-existent, and are construed as the exclusive domain of mining companies. One telling exemplar of this management mentality relates to recent minerals policy revolving around the in-situ leach Beverley uranium mine in South Australia. In early 1999, Environment Australia approved the US nuclear company General Atomic's environmental impact statement application to use an acid-based leaching process to extract uranium from the ore body. Once the 'commercially viable' uranium products are removed from the earth, the by-products are pumped back into the underground aquifer. Yet this process is illegal in the USA. There only alkaline solutions are used, and the by-products from the extraction processes are collated on the surface then disposed of in storage facilities. In Australia, terrestrial environments 'beneath the surface' are not deemed important. This promotes cheap and environmentally degrading disposal techniques. This mentality provides mining interests virtually unlimited access to conservation zones and parks, as directional drilling can be pursued diagonally underneath these zones without unduly upsetting the surface biota, and thus 'keeping the greenies happy'. Aquatic environments, on the other hand, are often seen under wise use as three-dimensional domains, for different reasons, but for the same purpose: to provide resource extraction industries with unlimited access.

An excellent example of the absence of 'no-take zones' in wise use natural resource management occurs in the conservative government's declaration of the Great Australian Bight Marine Park (GABMP) in 1998. The Great Australian Bight Marine Park is divided using sets of latitudinal and longitudinal lines, as well as horizontal and vertical ones. Two key management zones are proposed: the Marine Mammal Zone, and the Benthic (sea floor) Zone. Basically this translates into certain practices being pursued in different surfaces areas, as well as at different depths. This means that activities such as mid-water trawling, demersal shark netting and petroleum/gas extraction are allowed above and below the benthos, although these practices have obvious and significant impacts on the seabed. To add to this industry friendly regime, there aren't any no-take zones in the overall 'park'. Michelle Grady (1998, p 13) writes:

> Without any 'no-take' sanctuary zones in the Commonwealth waters, there are no control sites by which to assess the effects of trawling, fishing (shark and lobster), mineral and petroleum extraction. Hence, the proposed research and monitoring program for the park will not be able to demonstrate the effects/benefits of the zones—as the zones do not actually prohibit any activities.

Environment Minister Hill has argued that 'no-take zones' will no longer appear in his government's resource management strategies. In an ABC radio broadcast from Port Lincoln in September 1998, Senator Hill addressed the management of the Great Australian Bight. He stated that 'MPAs [Marine Protected Areas] which exclude use are an old-fashioned view ...' (Hill 1998b). GABMP is not a park at all, but an industry free-for-all, using the language of conservation as window dressing. This 'park' will prove virtually useless in protecting endangered species such as whales, dogfish and Southern Ocean tuna, and will fail to provide Australia with a genuinely sustainable fishing industry. This complex map of smoke and mirrors promotes all and denounces nothing: it is wise use at it destructive best.

ROUND-TABLE DECISION-MAKING

The wise use movement has been assiduous in its sale of round-table, consensus decision-making. The round table itself represents the imagined non-conflictive notion of pluralist political models, which promotes the win-win models of unfettered markets. A fine example of this championing of non-conflictive round tables is the Quincy Library Group. The town of Quincy, in the Sierra Nevada range in California, formed this group in 1993 ostensibly to resolve resource conflicts between timber and environmental interests, as well as government agencies (Sierra Club 1997, p 1). Meetings were held in the town library so people would not yell at each other. Wise use and multiple use models are based on the premise that everyone will win in the market, including the environment, if quiet negotiation between conflicting interests and values takes place. The Quincy Library Group has been extremely successful in promoting its wise use round table, resulting in victories to resource extractive industries. It has been hailed as a model of resource decision-making right across the USA. The Quincy Library Group Forest Recovery and Economic Stability Act of 1997 emerged from this process. It orders the US Secretary of Agriculture to hold a five-year pilot project based on Quincy-style decision-making on three national forests in the Sierra.

Although it appears to uphold pluralist notions of democracy, the round table is deeply exclusive. All decisions are handed over to 'local' people. The definition of local is telling, as it includes multinational companies and their management teams (as they are employers in the local community), but it excludes representation from national environmental organisations. Consequently, industry is markedly over-represented, creates agendas and bottom lines, sets terms of reference, and receives acclaim for achieving community consensus, with all the legitimising imprimaturs associated with this process. The notion of 'localism', of course, fits in nicely with the discourses relating to 'bioregions', 'grassroots', and 'participative decision-making': concepts widely endorsed by environmentalists. This further diffuses opposition. Sierra Club chair Michael McCloskey's November 1995 memo to the club's board comments on this cleverly contrived exclusionist process:

> Industry thinks its odds are better in these forums ... It has ways to generate pressures on communities where it is strong, which it doesn't have at the national level ... This re-distribution of power is designed to disempower our (national environmental) constituency, which is heavily urban ... Big business has a game-plan of pursuing this approach to get out of the clutches of the tough federal agencies ... A lot of people on the left have been taken in because it is a touchy, feely approach that plays to those with romantic notions about localism and self-control. They forget they're disempowering most of the people who have a stake in the issue. (McCloskey quoted in Mazza 1997, p 3)

Consensus is more easily achieved by limiting the domain of stakeholders. These round tables are also formed by business, not the state (although sometimes it is a shared task, with the division between private and public sectors almost imperceptible). The pluralist and even corporatist notions of state-initiated coordination, mediation, and monitoring of such processes are no longer useful in understanding these recent innovations.

The work of wise-user Patrick Moore is of interest here. Moore is North American and an ex-Greenpeace professional now working for industry. In 1998 Moore visited Australia, a guest of the Forest Protection Society, an Australian wise-use-style front group for the National Association of Forest Industries (Strong 1998, p 3). Moore was vociferous in opposing the 'locking up' of forests in 'no-go' zones in national parks and reserves. In his web site entitled 'Green Spirit', Moore coopts green and democratic language to further industry's gains. He gives much attention to successful round-table decision-making and offers advice on achieving agreement on membership and terms of reference. Moore (1997, p 3) describes the purpose of these round tables:

The job of the facilitator, in the final analysis, is to help the round table produce a consensus document, which expresses the areas of unanimous agreement among the participants, and where there is not unanimous agreement, an explanation of that disagreement, in words that are unanimously agreed to by all the participants.

Win-win, non-conflictive decision-making imagines government in limited economic terms. This neo-liberalism and radical libertarianism sits neatly in, and builds upon, the earlier discourse of sustainable development, which argues that all environmental problems can be resolved by increasing efficiencies at the political, social, economic, scientific and organisational levels (Doyle and McEachern 1998). Thus all environmental problems can be resolved with better management, they just need to be addressed using professional conflict resolution techniques. Jim Britell (1997, p 6) indicates the shortcomings of such an attitude when he argues: 'A defining characteristic of this approach is a reluctance to ascribe the cause of any problem to pervasive and systemic corruption, or to the ability of the rich and the strong to take advantage of the poor and the weak.'

In fact these round tables are exclusive, they are based on narrow terms of reference, and they operate from a restricted and biased information base. At a symbolic level 'they remain a corporatist exercise designed for quieting opposition and coopting more radical, non-governmental strategies' (Doyle 1998b). They deliberately ignore power differentials between stakeholders, while 'manufacturing nature' in economic rationalist terms.

Wise use and sustainable development work hand in hand with free-market economics. As the market is deemed 'natural', the ecology of the ecosphere becomes 'the market'. All inputs and outputs are given value in monetary terms and then, so it is argued, the 'natural', 'real' and 'essentialist' economy of ecology will emerge, unfettered by the constraints of science and state regulation. The 'trickle down effect' will benefit those species on the lower rungs of the natural hierarchy, promoting widespread ecological health and doing away with any notions of science-generated ecological safety nets thrown over the most disadvantaged, those species and habitats most at risk. Nothing is irreversible, everyone and everything will win: the chocolate ration has been increased. This line of argument, of course, also fits neatly into the parameters of neo-Spencerism, promoting the notion that those human and non-human communities most likely to survive unfettered 'natural' systems will be all the better for doing so, having weeded out the weak.

In the Australian context, this ideology has led to pressure being placed on ecological safety zones to produce economic profits: 'parks must pay'. This logic is similar to that shown in the development of overseas aid programs that must show profitable returns for the donor country. The state will no longer fund, provide and support its own management regimes.

THE ATTACK UPON THE PUBLIC SERVICE AND 'INDEPENDENT SCIENCE'

The state is retreating from past responsibilities, it is no longer the hub of the round table. There is a close collusion between the state and big business. The state now presents itself as a corporation. Recently in South Australia, both the Department of Primary Industry (PIRSA) and the Department of Environment, Housing and Aboriginal Affairs (DEHAA) have undergone a structural transition to a purchaser/provider model, which has been echoed in numerous other bureaucracies under free-market governments elsewhere in Australia and the world. The purchaser/provider model dictates that the purchaser, or client, is no longer the public. Each bureaucratic division's client is now its own departmental director and his or her policy advisers. The director purchases the services from his or her own employees. When these services can be more efficiently and effectively purchased outside the bureaucracy, the director outsources them. Where once the public would demand that the public service act in its interests, now the bureaucracy must satisfy its own CEOs, who, in South Australia, are sometimes political appointees of a party whose interests are often inseparable from those of big business. The bureaucracy must now provide the 'correct' information that will lead to profitable outcomes. In this manner, business becomes the client and government becomes its service provider.

One critical impact of this situation is the loss of public input that once led to the state acting on behalf of citizens in the processes of mediation, data generation, monitoring, legislation and regulation. Non-government organisations are now often forced to bypass the state and attempt to negotiate directly with powerful companies. Understandably, there is no pressure on the company to engage in private negotiations with environmentalists, as it has no responsibilities to the general public; rather, its chief concerns are its shareholders.

With this attack on bureaucracy under wise use, there has also been a thorough offensive against 'independent science'. This next step is a logical one for the coalition government and big business, as bureaucracy is the most 'scientific' form of governance, based on seniority, merit and objectivity. In many ways independent science parallels the concept of a separate, permanent state, performing a monitoring and moderating role in the processes of capital production, accumulation and consumption. Whatever the limitations of bureaucracy and positivist science, both processes provide an alternative force to the excesses of advanced global capitalism. As an autonomous state is often seen as interfering in radical free-market environmental solutions under sustainable use (and the more extreme variant, wise use), bureaucratic and other forms of 'independent' science may also be a barrier to market-based outcomes.

The funding and political support for the CSIRO, for instance, has been dramatically slashed in real terms under Howard's neo-liberal economic regime. Generally, however, the attack on science is more subtle, masked in the changes to the philosophy of government already alluded to in the discussion of the purchaser/provider model of the public service. Fewer permanent scientific officers are being employed, with more and more of these services being outsourced to industry-friendly consultants.

A fine archetype of this process is found in the South Australian public service. Originally, scientific officers were hired by the Native Vegetation Assessment Branch of DEHAA to independently assess applications for native vegetation clearance. Over recent years these permanent employees have been replaced by temporary employees provided by a 'temp. agency'. The permanent employees of the state had infrastructural and institutional support to embark upon research and to make recommendations without fear or favour. It is fairly obvious that the careers of temporary officers are far more susceptible to political manipulation. As of April 1999, of the twelve scientific assessment officers in the branch, there were only three full-time and one part-time permanent scientific officers left.

Even more disturbing is a recent trend in the same branch. The agency has now been restructured as an environmental consultancy under the name EAC. This 'consultancy' has performed its own 'rapid assessments'. For an additional cost, certain companies can have the normal bureaucratic assessment procedures fast-tracked, delivering the proponent, more often than not, market-based (favourable) outcomes. Mildara Blass, an Australian wine-making giant, is one company to pursue rapid assessment in this way. The boundaries of public service and private consultancy are blurred here, as are those between independent and market-led science/non-science. Outsourcing becomes insourcing, and incipient corruption becomes a greater possibility.

Whatever the political rhetoric, this situation, coupled with a weak Native Vegetation Act and an inactive state Environment Minister—Dorothy Kotz—has led to a dramatic increase in both legal and illegal native vegetation clearance in South Australia. In a news release in April 1998 the CCSA (1998b) stated:

> In March, 1495 trees and 19 ha of native vegetation were recommended for clearance approval ... Large applications are currently going through

the system, applications for high value vegetation which include old trees which would take hundreds of years to replace. The Native Vegetation Council (the statutory body which administers the Act) seems powerless to stop this endless stream of destruction.

This situation is particular significant in a state where in some areas as little as 2 per cent of native vegetation remains.

The outsourcing of science continues across the board. In the Simpson Desert Regional Reserve Ten Year Review, the SA government 'let go' the review to a private consultant, Resource Monitoring and Planning Pty Ltd. The consultant requested any 'information regarding relevant activities of the numerous stakeholders in the area over the last 10 years, and some basic descriptive background material' (RMP 1998). In a response to this management/public relations company, Vera Hughes (1998), state coordinator of The Wilderness Society, writes:

> The ecological quality of the Simpson Desert Regional Reserve over the past ten years is impossible to assess given the lack of scientific data currently available. We simply have no credible means of measuring the impact of exploitation. From your letter, it is clear the focus of your concern is on gathering this data any way you can. Basing studies on data provided by vested interests may be better than nothing. Or, it may not. If the attempt to assess the viability of the regional reserve for conservation is sincere, ongoing scientific data must be gathered and interpreted by independent parties. If this data has not been gathered on an on-going basis, a thorough and rigorous audit of the Reserve is not possible.

When the state relinquishes the responsibility of providing or adequately funding the compilation of base-line scientific data upon which to establish a management regime, each stakeholder whistles in the dark.

Laissez-faire ecology is a nonsense and non-science. It does not acknowledge the trend in biology away from Darwinist competition models. Instead, it nicely supplements the politics of neo-liberalism and the globalisation of advanced capitalism. Within it, there is no understanding of the reality of relationships within and between societies and ecosystems. To consider nature as an ecological free-market will lead to massive species extinction and habitat degradation, while also extending inequities between human beings. Unrestrained commodification of the planet Earth will not lead to win-win games, but, in the short term, to a few large wins for a powerful minority; while in the long term it will lead to massive losses for all forms of life on the planet.

We should note too the inadequacy of several public-policy texts in their analysis of resource conflicts under conservative Australian governments. Many of these texts, such as Bridgeman and Davis's *Australian Policy Handbook* (1998), still articulate the central and separate role of the state in resource decision-making, almost ignoring the invasive power of business, while providing an ideological framework in which wise-use-style, win-win round tables derive substantial credibility. These outdated, rational comprehensive-style texts operate on the widely accepted but unquestioned premise that the '*model of society* underlying the contemporary rationality project is the market'. Stone (1988, p 9) critiques this position:

> Society is viewed as a collection of autonomous, rational decision makers who have no community life. Their interactions consist entirely of trading with one another to maximise their individual well-being. They each have objectives and preferences, they each compare alternative ways of attaining their objectives, and they each choose the way that yields the most satisfaction. The market model and the rational decision-making model are thus very closely related.

The politics of wise use, in its brutal depiction of nature as marketplace, coupled with the decline of the central resource management role of the state, has driven environmentalists to the edge. Although some of the corporatist strategies of the second period remain, these negotiating round tables are now increasingly defined and dominated by business. The change is indicative of other changes in the character of the state in Australia. With these political changes and the impact on our lives of globalisation, the role and identity of the nation-state is now more confused than ever. New models of power are sometimes needed to describe this political milieu.

POSTMODERNIST AND POSTSTRUCTURALIST DESCRIPTIONS OF POWER, SOCIETY AND THE STATE

Postmodernism and poststructuralism are terms under which a multitude of different forms of academic theorising has developed. In different disciplines, the impact of the postmodern and poststructural is quite different. Elaine Stratford (1995, p 13) writes:

> In architecture, literary criticism and theory, literature and the arts, postmodernity has been used long enough for some observers to suggest that the disciplines have moved beyond the concerns of postmodernism. In this instance, postmodernism is defined as a set of practices which critique the modern turn, whose inception is dated anywhere from the Renaissance to the middle of the last century. However, in the social sciences, although the label postmodern is used, the term poststructural also describes disparate works that critique foundational assumptions on which the knowledge of modernity rests. Among the 'sacred cows'

targeted by poststructuralism are the ideas that the body is an essential object, the (Cartesian) self is separate from the body in which it is located, that there are universal and immutable truths from which we can derive ideologies, and that—among others—nature and culture, the masculine and the feminine, the public and the private are binary opposites.

Several of these key elements of postmodernist critique against positivist Western thought were presented in the introduction as premises or myths underpinning this research.

Poststructuralist critique in the social sciences emerged from the writings of a network of French social theorists, led by Michel Foucault and Jacques Derrida. For the purposes of this brief exposition, however, poststructuralism is understood here as part of postmodern analysis. I use the two terms loosely and interchangeably here. Postmodernism is often used to provide sustained critique of the project of modernity. The social project of the Enlightenment, and the myths of rational science, progress and an instrumental nature (among others) that are attached to it, are seen by some postmodernists as central to the concept of modernity. Postmodernists challenge the modernist world view, and on occasions, this stance also becomes anti-modernist, since they define their position in terms of a rejection of modernity. Sometimes the term postmodernism is also used as a synonym for the advanced modern or the *hyper*modern.

Postmodernist theory regards language as extremely important. Roger Sibeon (1996, pp 12–13) writes:

> Post-modernists along with poststructuralists and ethnomethodologists believe that discourses constitute or 'construct' the social context. The social world is not preconstituted. There is no real, pre-existing social reality 'out there': there are only languages and discourses (the latter as sets of linked meanings and practices) which construct 'reality' that is experienced by the language-user.

Postmodernist theory rejects the structuralist arguments of 'one-way causal determination' (that is, for example, that institutions determine, to a large extent, the actions of people), arguing that language or discourse moulds the sociopolitical context as much as the social context shapes the discourse. This idea is anathema to structuralists, but it is much more than just an attack on structuralism: it is a broadside on Western metaphysical thought. It questions the construction of what Stratford referred to as binary opposites. White argues: 'The first part of the opposition delimits the sphere of what is privileged and foundational for the construction of a secure political world, the second half, the sphere of what is marginal, suspicious, and ultimately seditious—what poststructuralists often refer to as the "Other"' (White 1991, pp 15–16).

Postmodern theorists contend that strict divisions between categories such as men and women, the bourgeoisie and the proletariat, or the 'haves' and the 'have nots' are not solely historically determined categories, but are often the result of actors' discourses: they are 'mobile categories, formed to fit the conditions of their use and user' (Fiske 1993, p 6). Although these epistemological and ontological positions are sometimes exclusively argued in 'high theory' circles, they do actually relate to key issues of society, the state, and power.

Social commentators increasingly argue that advanced industrial countries such as Australia are experiencing postmodern or hyper-modern characteristics such as social diversity, cultural pluralism, post-industrialism and consumerism. Sibeon (1996, p 12) describes some of these characteristics as follows:

> Diversity, heterogeneity, and pluralism are said to be the crucial defining features of the post-modern. Instead of simpler class-based divisions of modern society, diversities in post-modern society are linked to multiple and cross-cutting values and affiliations to do with gender, race and ethnicity, religion, sexuality, generation, regionalism, and localism. Alongside this social fragmentation and the breakdown of the 'grand narrative' is an extreme cultural pluralism, so that post-modern societies are said to be a babble of ... mutually uncomprehending discourses, lacking a coherent ideology or central authority.

Postmodern models can be used to describe Australian society in the late 1990s. There is some debate, however, as to whether Australia qualifies as a first-, second- or third-world country. Its economic system, which is so reliant on minerals exploitation and other primary industries (with little value added) is strongly reminiscent of many of the neo-colonial economies to its immediate north. Whatever Australia's economic status, it has a large, white middle class that is still dominated by European culture, despite the population's changing racial mix. In Australia this sizeable middle class enjoys a relatively high standard of living when compared to many nations in the South. Australia is one of the most urbanised countries on the globe, with nearly half of its 18 million people living in Sydney or Melbourne. Most Australians see their country as post-industrial and part of the First World.

Unlike parts of Europe where there have been some substantial adjustments made to laissez-faire economics and practices over the past decade, Australia (chiefly under conservative state and federal governments) has operated on radical libertarian market principles, experiencing Thatcherism and Reaganomics all at once, but lacking an intellectual tradition to provide any sustainable challenge, or to provide any subtle edges to these hardline, antisocial approaches. Australia is increasingly moving away from much of its cultural and economic association with the United Kingdom, and is now more often seen in the South as the junior partner of the USA in the Asia–Pacific theatre.

This situation may not have occurred as a result of post-industrialism/postmodernism, but simply because post-invasion Australia has only existed for little more than 200 years and, apart from its indigenous culture, has always possessed the features of social fragmentation and ideological incoherence. These characteristics have been exacerbated in recent times. In a global sense, to an extent white Australia has always been postmodern: the flotsam and jetsam of industrial Europe were either driven through necessity or attracted by new possibilities to a 'new world'.

Some argue that in poststructural/postmodern societies there are a multitude of conflicting systems of meaning and value. Technology, it is contended, has globalised our societies, leading to the end of tribalism and isolation, and to an increased tolerance and acceptance of cultural difference. The metanarrative has ceased to exist, replaced by smaller, more relativistic and particularistic narratives (Dickens 1990, p 105). At first glance, these descriptions seem quite apt in the Australian context, but it is not clear what has caused this situation. There has been a profound dissolution of the concept of community over the past generation, with fewer urbanised Australians knowing or dealing with the 'people who live on the other side of the suburban fence'. There has been a dramatic retreat from organised religion and other organisations offering a shared meaning.

Sibeon's description of a postmodern society also matches, in many ways, the smeared-out conceptualisation of the environment movement presented in chapter 1. Importantly, however, a movement is not a society, as a movement is not a formal organisation or an informal group. Societies and social movements are very different human collective forms, with different traits and characteristics. Having said this, my depiction of the movement as lacking unified goals and ideology originally emerged in isolation from the literature and debates of the postmodernists. The science of ecology, with its myths of interrelatedness and connectedness, coupled with attempts to understand the informal politics of movements, had more to do with the three-dimensional, dispersed and fluid depiction of the movement that I have termed the palimpsest. Of course, this can easily be explained away in the literature of the postmoderns, with less emphasis given to the creative powers of the 'author', who is only responding within and being informed by dominant discourses, regardless of conscious deliberation.

Postmodernism also constructs both the state and big business very differently to more traditional models of political analysis. White (1992, pp 17–18) contends:

In one sense, Foucault would agree that contemporary Western society has seen a diminution of 'state power', when by that term we mean the arbitrary and repressive employment of mechanisms of coercion. His key point, however, is that we must understand power also in another way, namely, as a slowly spreading net of normalisation that invades our language, our institutions, and even (especially) our consciousness of ourselves as subjects.

Power, then, is 'exercised from innumerable points, in the interplay of non-egalitarian and mobile relations' (Foucault 1976, p 94). Due to a view of power 'as a shifting and precariously sustained outcome of networks and alliances and of the operation of discourses' (Sibeon 1996, p 115), the state is not perceived as a unitary actor. A poststructuralist position similarly sees business interests as neither unified nor necessarily explicit. Distinctions between the interests of the business community, environmentalists and other sectors, including the state, are now increasingly blurred. The categories of 'the other' are interspersed and fluid: there are no 'enemies'. Finally, it is suggested that power does not necessarily have to clash in dualistic confrontations. Rather, it is multidirectional, dispersed throughout society.

In depicting power within societies as blurred and dispersed, we need to realise that not all sectors are equally dispersed. In this third period, for example, with the state reneging on many of its responsibilities, power has become more centralised in the business sector. Although not absolutely uniform or united, business interests have gained power in recent times, and it is being repeatedly used to directly threaten dissenting interests within our society. To accept the palimpsest model of the movement does not mean that we view all sectors in society as exerting power in a similarly dispersed and uncoordinated fashion. Indeed, the smeared-out, non-ideologically unified vision of the movement may have less to do with the nature of postmodern societies, with their supposed myriad discourses and choices, than with a survival mechanism, which allows the movement some medium- to long-term resilience, enabling it to endure the broadside levelled at it by big business. Thus the so-called postmodern form of the movement may not result from a multiplicity of positive effects (now that the meta-narrative Enlightenment goals of modernity have been put to rest), but rather as a result of having to duck behind every nook and cranny to survive the onslaught of the grand narrative of advanced capitalism, which has become so powerful that we are told it does not exist.

The best critiques of poststructuralism emanate, not surprisingly, from the structuralists. Many theoretical activists in the South, for example, are quick to denigrate academic works depicting the globe as a 'postmodern space' as Northern, neo-liberal propaganda. Jose Maria Sison (1998, p 3), a respected movement intellectual in the Philippines, writes from this angle when he states:

The most outlandish claim of the 'neo-liberals' is that multinational firms and banks have lost their national character and basing, and have become so powerful internationally as to render impotent all kinds of states and reduce to irrelevance all questions of national sovereignty. In fact, multinational enterprises have most of their capital, stock owners, personnel, research and development in their home countries and depend on their own imperialist states and multilateral agencies of states for protection, insurance and subsidies in their overseas operations and expansion.

As I argued in the preface, in many nations of the South, structuralist theories (sometimes slightly reconstructed) still inform the majority of the politics of resistance: in the churches, in the cities and villages, and in human rights, indigenous peoples' and environmental NGOs. Although the majority world has been under sustained attack from the North for over a century, in many parts of the South there has been no dissolution of the grand narrative. If anything, definable oppression has increased, particularly in the last decade or so, in part as a result of globalisation and the concomitant increases in the power and freedom of transnational corporations.

Some of the more traditional critical analysts would perceive this, for example, as a deliberate, instrumental power play on the part of major business brokers, and others of a more structuralist bent would say that this is the just nature of processes and routines involved in the production of capital. Structuralist analysis does not depict the state as neutral (as the pluralists would have it) or unidentifiable (as some post-modernists would contend), but rather an instrument of the ruling elite to promote its own interests. Within this structuralist view, the interests of the state, capital (whether national or transnational) and elites are largely indistinguishable, so that any incremental changes to the laws and legislation that serve the state are perceived as only further enhancing its legitimacy and authority.

* * *

Although the ALP under Hawke corporatised environmental concern to such an extent that it became part of the mainstream furniture, the Howard government has moved to a more rejectionist stance on environmental issues that is, at first glance, reminiscent of the first period, when greens were perceived as outsiders and enemies of the state (despite their predominantly pluralist politics). Unfortunately for the environment movement, it is far worse than the first period, when modern concern about the environment was being forged into a political force. Now the movement's strategies and tactics are largely a known quantity, and the public relations industry's opposition to green political theory and activism is now extremely sophisticated, supported by large amounts of money emanating from the private sector.

The round tables forged under the Labor government's Accord, using the corporatist models of policy-making, are now simply window-dressing. The state is largely a participant in these round tables, rather than the central provider. In this third period, access to corporatist round tables is limited to those who are willing to accept money from corporations, who are willing to sell their integrity. In many ways, to talk of round tables is now misleading. Round tables normally denote the participation of more than two players. More and more, politics under the Howard government and in conservative-led state governments is simply a direct partnership between two players: business and the state, with little delineation between the two.

The environment movement has been driven between a rock and a hard place during this era. To its credit, it has still survived as one of the few social forces that critiques business-as-usual, but it is struggling, its resources are spread membrane-thin across a plethora of campaigns. Despite the negative reporting of Murdoch- and Packer-owned media outlets, there are more issues now on the movement's agenda than ever before. We examine in the next chapter the strategic direction of the movement in this most troubled and oppressive era, to some extent facilitated by the corporatist-style, assimilationist policy-making practices of the second period under Labor.

NOTES

1   This analysis of the wise use ideology appears in full as a chapter in KJ Walker and K Crowley (eds), *Environmental Policy Two*, University of New South Wales Press, Sydney, 1999.
2   The author spent six months researching the emergence of American right-wing social movements at the University of Montana in 1996. Much of the analysis of the wise use movement's environmental strategies included in this chapter emerged during this period. Along with the works later alluded to by Dowie and Rowell, other excellent wise use critiques are found in D Helvarg, *The War Against the Greens: The 'Wise Use' Movement, the New Right, and Anti-Environmental Violence*, Sierra Club Books, San Francisco, 1994. In Australia, some of the strategies of the wise use movement have been ably monitored and published in a number of articles by Bob Burton, for example, 'Right Wing Think Tanks', in T Doyle (ed), *Chain Reaction*, special edition 'Corporations, Power and the Environment', no. 73–74, 1995.
3   Of course, 'stakeholders' cannot be confused with 'people' or 'the public'. The concept of the stakeholder is often used to exclude any further dealings with 'the public'. The choice of the appropriate stakeholders is an intensely political process that is rarely democratic or representative of broader interests.

PLAYING POSTMODERN GAMES?
THE POLITICS OF RESILIENCE

To describe the movement in postmodern and poststructural terms is a political choice that leads to different analytical and strategic outcomes. The nation-state has been weakened by the growing power and acceptance of the politics of globalisation with its culture of free markets. This is government as decried by the Dow-Jones index: the power of elected representatives of the people is greatly diminished in this context of dispersed, multitudinous non-centres of power.

Is this supposed postmodern political space the result of tolerance and the new politics of global choice, or has capitalism won? Have grand narratives really disintegrated or has one grand narrative become so powerful in postmodern societies that it has vanquished alternative theories while managing to sell the concept that a whole variety of conflicting points of opposition exist, reflecting a healthy society? It may be accurate to depict the movement in Australia as a non-goal specific, smeared-out palimpsest when non-institutional, green politics in society (or non-society) is now so overrun by a single metanarrative that all opposition has been dispersed and cast as disparate 'minority voices'. It is virtually impossible to single out this metanarrative, as the story actively sells the concept of its own non-existence. As part of this brilliant disappearing act, citizens are told that they live in a 'pluralistic' society with a multitude of healthy, conflicting ideas and belief systems, that 'democracy' is alive and well, and that mass opposition is not warranted. So powerful is this grand narrative that it is not popularly perceived as a discourse or theory at all, but the market simply is deemed to operate on 'natural' properties and truths. In postmodern states such as Australia, incremental reform within advanced capitalism is imagined, while radical and revolutionary changes are rarely articulated. So dominant is Mammon that 'capitalism' is never mentioned in the polite company of the chattering classes, who are silently acquiescing to the new wealth-accumulating plutocracy. With the commodification of all the Earth and its people, citizens are now individualised consumers and individually consumed, with any notion of society as a composite entity now gone. These processes of capital accumulation have quickened, with the inevitable outcome being further discrepancies between the rich and the poor.

Yet during the Howard years, the Commonwealth and the state governments cannot simply be perceived as victims of 'greater forces'. The Commonwealth has abdicated its responsibilities in social and environmental spheres, while promoting the interests of business. The raft of environmental legislation built in the first flush of environmental activism in this country was celebrated all over the world. This surge of legislation, created by both pluralist and structuralist forms of environmental opposition, stopped to a trickle during the second period under Hawke. In this third period, this legislation has been gutted.

Whatever the criticisms of environmental policy-making under Hawke, in a number of landmark legal decisions (including the Franklin), the Commonwealth took on board the chief responsibility towards environmental protection. The conservatives, with their traditional power base more secure in the states, have worked assiduously to reverse this trend. Under the National Environmental Protection and Biodiversity Conservation Act, the Australian states, which have been prime movers in many of the development projects that have raised most environmental concerns, have now been given the power to resolve environmental conflict. Environment ministers are now principally responsible for 'keeping the lid on' environmental concerns, rather than actively addressing them. With the low profile of the state, the ideology of the global market and its resource management dogma of wise use achieves greater influence.

The question in this chapter, then, is where do environmentalists go, what strategies do they pursue in this apparently postmodern political space? In the first period, activists playing pluralist games sought indirectly to influence the state to provide better legislation, to become a better mediator and regulator in relation to environmental management. Structuralists, particularly in the anti-nuclear movement, pursued more grassroots strategies, convinced that the massive groundswell of educated and empowered citizens would be enough to achieve greater societal changes. Both strategies were outside the state, but at least there *was* a state to either indirectly influence or directly demonise. In the second period, large parts of the movement were at times incorporated into the state. Many aspects of environmentalism became mainstream: institutions and business incorporated these concerns into their agendas.

During this third period there has been a resurgence in the anti-nuclear movement. While wilderness networks and organisations have grown older and fewer, the anti-nuclear networks have blossomed, attracting large numbers of younger activists. There are a number of reasons for the retreat of the wilderness networks. First, the concept of wilderness is increasingly seen as an uncomfortable one in relation to indigenous Australia. Second, wilderness activists have not managed to include environmental justice equations, such as equitable resource distribution to all people, on their agenda. Third, their enthusiastic embrace of corporatist politics under Hawke has weakened the more radical networks. Fourth, many networks are now being punished for their close association with Labor. Those that have retained or gained access to government are only involved in window-dressing: receiving vast amounts of money and kudos from Robert Hill, via the mining and pastoral sectors (among others) to plant trees and to 'save' species.

Fifth, in its desperate bid to avoid being labelled an 'anti-development' party, the ALP has largely dumped concern for the environment from its agenda and has provided no sustained criticism of the coalition's lack of environmental governance. As we saw in chapter 6, in the 1998 federal election, Labor informed the green elite that did it not want the green vote. Some mainstream green activists have continued to dabble with parties, but with a discernible lack of conviction: this is not where the movement has spent its time and resources. Finally, most wilderness networks have ceased attempting to influence law-makers and have become involved in 'individual value change' (more on this later).

Anti-nuclear networks have re-emerged in the 1990s as more viable and oppositional voices in Australian society, channelling the energies and aspirations of Australia's youth, while the wilderness-oriented 'greenies' age. Anti-nuclear networks possess broader social goals, which provide a greater sense of relevance to younger people, and they have maintained their more structuralist strategies of mobilisation through grassroots empowerment. For these and other reasons, such networks are more capable of playing politics under a jackboot. Yet both major parts of the movement are struggling, both are seeking new strategies and revisiting old ones.

In this third period, what can activists do when the state fails to provide any real response to its concerns, when it is not involved in genuine and substantive legislative change, and when it abdicates its primary responsibilities to the point where elected politicians cannot even be called to account? Removed from the power of the state and having lost the capacity to influence changes in legislation and policy, the response of green groups, NGOs and informal networks that have survived has been varied, though some major trends have emerged. The movement needs to distinguish between diverse, resilient strategies and the neo-liberal politics of confusion, which can only deliver outcomes to many of its traditional adversaries.

It is the transition from *descriptive* poststructural models of power and policy-making to *prescriptive* ones that is most problematical for the environment movement. In numerous local, state and national green movement think-tanks and introspective gatherings in the late 1990s, movement strategists constantly reaffirmed that Australians, as global citizens and consumers, were now living and working in a postmodern, globalised space. Unfortunately, whatever the limited merits of this description, the argument that 'the political system is now postmodern, therefore our strategies must also be postmodern' does not necessarily follow. This viewpoint is, however, advocated by certain academic theorists as well. Postmodernist depictions of society, outlined in the previous chapter, have been seized upon by numerous 'new social movement' theorists. Many of these theorists, in accepting that societies have become postmodern, prescribe postmodern tactics and strategies as a way of playing politics in this new political sphere. White (1991, p 11) writes from this perspective when he contends:

> What stands out about new social movements is an irreducible pluralism and a suspicion of totalistic revolutionary programs. For such groups, the growth of incredulity toward traditional metanarratives is particularly unlikely to be seen as a source of fear and trembling, but instead as a source of some cognitive space within which new orientations may have a better chance to flourish. This is especially clear in the case of the women's movement and radical ecologists.

One consequence of this denial of the existence of metanarratives of dominance or resistance is panic and confusion. This often manifests itself as 'the scatter-gun approach', which argues that the movement should pursue all possible, even limited avenues open to it at all times. Description of these strategies is difficult as the list is a long one. I have chosen examples for their diversity and their interest, and have attempted to be representative. Despite this heterogeneity, the other major outcome of this period has been an increasing tendency to bypass the state.

WILDERNESS AND ANTI-NUCLEAR CAMPAIGNS DURING THE THIRD PERIOD: BYPASSING THE STATE

Figure 12.1 illustrates this new political situation for both the anti-nuclear and wilderness networks. It shows the state no longer at the centre. The state is most often seen as an additional but often silent stakeholder, allowing its more vociferous partner, big business, to make the running on key environmental issues and, for example, to set industry operational standards.[1] Key individuals within the movement do still deal directly with ministers and their opposition counterparts, with members of what once was the permanent arm of government—the bureaucracy—and with party officials (usually from the minority parties such as the Greens and the Democrats). We turn now to three different forms of strategic politics applied during this period: playing corporate games, pursuing individual value change and building a global soul, and organising minority and mass mobilisations.

Figure 12.1
Postmodernist pathways in wilderness and anti-nuclear campaigns during the third period

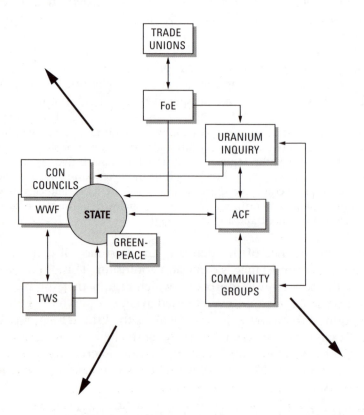

Note: Green groups in the third period no longer look exclusively to the state to achieve environmental goals, so here it is not represented as being at the centre of the process.
Compiled by the author.

Postmodern elements could be used to describe each strategic pathway. The first, working directly with and against corporations, could be construed as a result of globalisation, with the dominance of the market over nation-states forcing activists to bypass the state. Next, with the perceived acceptance of this new globalised space, there has been a growing trend towards refashioning the *values* of global citizens (or consumers) to provide what has been termed a 'global soul' for advanced capitalism. Finally, although mass-mobilisation campaigns still occur (despite the attempts of globalists to denigrate the concept of community as tribalism), they are more event-oriented and less community-based than similar campaigns in the first period.

PLAYING CORPORATE GAMES

Even in the movement's own publications, the term 'lobbying', which once usually meant the activity of indirectly influencing government in the foyers of parliaments, now refers to influencing corporate activities. In describing the multifarious strands of the anti-nuclear campaign to stop the Jabiluka uranium mine in the late 1990s in *Green Left Weekly*, Malloch and Fulcher (1998, p 9) describe 'lobbying' as follows:

> Lobbying: Almost since the campaign against Jabiluka began more than two years ago, a 'corporate campaign' has been led primarily by the Wilderness Society and activists close to it, including the Minerals Policy Institute. The basic content is lobbying two corporations associated with Energy Resources Australia—parent company North Ltd and Westpac bank—to cut ties with ERA because of its involvement in the Jabiluka mine ... North Ltd is also being targeted with a 'North Ethical Shareholders' campaign. It aims to convince shareholders that ERA is an unethical and unprofitable investment and to call an extraordinary general meeting to discuss Jabiluka.

In this description of lobbying there is no mention of government. These anti-business campaigns, resulting from the state's increasing absence, attempt to discredit certain corporate practices. The hope is no longer to change legislation, but to diminish consumer or investor confidence in certain companies, so that these 'environmentally unfriendly' companies might suffer in the consumer- and investor-oriented market place. Although most of these strategies are still appeals to corporate elites and their supporters, they remain radical when compared to a range of other strategies that simply accept the dominance of global business and seek incrementally to alter its practices from within. This latter set of strategies is more numerous and dominant.

In chapter 5, I argued that the World Wide Fund for Nature uses this more conservative appeals-to-corporate-elites strategy, which is premised on the motto: 'If you can't beat them, join them'. For example, in 1998, WWF endorsed Western Mining Corporation's 'Environmental Monitoring Report' with a glowing letter published in this report, which later led to WMC's receiving yet another industry award for environmental protection. Having served on the Olympic Dam Community Consultative Forum (CCF), I wrote to WWF querying why their 'report card' seemed to be at such variance with the on-the-ground experience of CCSA concerning the company's environmental performance. I received a reply from WWF explaining that it was only assessing the quality of corporate reporting, not the corporate practices. This is simply 'green-washing'. WMC is expanding its copper-uranium mine, which will be one of the largest uranium mines in the world.[2] Much opposition to the mine is based on anti-nuclear sentiments, and advocacy for the basic rights of Australian indigenous peoples. WMC's mismanagement of uranium mining and milling has been well documented and is not the focus of attention here, but it is worth noting that in February 1994, 5 million cubic metres of radioactive liquid leaked from the Olympic Dam tailings retention system. After this leak was reported, it was discovered that it had existed for at least a year without WMC bothering to make it public.

WWF representatives, and several other key environmentalists, who trade on the 'integrity' of their name for money and power, continue to be impressed with the company's efforts to look after the 'stick-nest rat' that is found in the vicinity of Olympic Dam, while the largest expansion of the uranium industry in this country's history continues unabated at Roxby Downs. Some anti-nuclear networks are involved in the Olympic Dam CCF process.[3] Although the odds are stacked even more against environmentalists than in the old government-led round-tables of the previous period, in this imagined postmodern space some activists argue that if they do not engage the corporation and the state here, then their chances of being involved in any decision-making process are almost non-existent. If they remain outside the process, others (such as WWF) willing to provide the 'green stamp' will be summoned. Another option—mass-mobilisation campaigns—has its own problems when the focus of the campaign, in this case the mine site, is not seen as aesthetically marketable (like Kakadu). In addition, it is increasingly difficult to achieve mass mobilisation, as it requires the existence of a community. In this postmodern space, with society as market and individuals as consumers, it is also difficult to initiate community action based on shared belief systems. Let us look briefly at corporate round tables, using the example of the CCF at Olympic Dam.

Although it appears to uphold pluralist notions of democracy, the CCF round table is deeply exclusive. The Olympic Dam Community Consultative Forum (CCF) is exclusive on four counts. First, all decisions are handed over to 'stakeholders' described as 'local' people. As

The launch of the Public Inquiry into Uranium. *Aussie Kanck*

we saw in chapter 11, 'local' here includes multi- and transnational companies and their management teams but excludes representation from national environmental organisations. Second, such round tables have an in-built bias towards the market, as the process reflects the neo-liberal politics of the marketplace.

Third, there are no representatives in the forum from the working population of Roxby Downs (over 1000 workers) other than managers. Both the Construction Forestry Mining and Energy Union (CFMEU) and the Australian Workers Union (AWU) operate on the site, but neither union nor non-union workers were asked to sit on the Community Consultative Forum. Fourth, and most important, is the virtual exclusion of Aborigines from the process. Only one indigenous representative was chosen from the community, and this was done without consultation. Responding to a letter from the CCSA in February 1998, the Minister for Primary Industries, Natural Resources and Regional Development, Rob Kerin, justified the inclusion of only one Aboriginal member on the CCF as follows:

> It was not easy to resolve the matter of an appropriate level of Aboriginal representation on the CCF. I am advised that there are at least seven different Aboriginal organisations with substantial interest in the matters to be discussed by the CCF. I believe it would be impractical to include representatives from all the organisations and choosing only two or three groups could offend those omitted. (Kerin 1998)

The minister resolved his dilemma by appointing Garnet Wilson, who was not a member of any of the claimant peoples involved with the mine sites and their bore fields. Wilson is a highly regarded Aboriginal leader and he was placed in an extremely difficult position, yet his inclusion as the sole Aboriginal member on the CCF was entirely inappropriate. How can it be a Community Consultative Forum when there are no Aboriginal members from the community? The minister was placed in a very difficult position due to the prior activities of WMC. It is not appropriate to discuss these issues here, as they are the business of the indigenous peoples themselves. It appears, however, that WMC has supported the establishment of small Aboriginal groups that have challenged the rights of the Kokotha and Arabunna peoples. Regardless of the manoeuvrings of WMC, in June 1994, 80 elders from various regions of central Australia met at Port Augusta and determined that the Arabunna people were the traditional custodians of the area in question. A committee of spokespeople and elders was nominated and confirmed in association with the South Australian government. WMC has not satisfactorily consulted with this committee. Having succeeded in confusing traditional ownership issues, the company used this confusion to justify non-consultation with 'local blacks'. As we saw in the preface, WMC employs similar tactics against indigenous people elsewhere.

The *Roxby Downs (Indenture Ratification) Act 1982* prohibits the publication of any material passed between the state government and WMC without their mutual consent. The implications of confidentiality arrangements are significant, to say the least. Shared secrets lead to further shared arrangements. Community groups that do attend these 'consultation meetings' often defend 'poor process' under wise use on the basis that they have been given singular access to important information. This argument does not apply in relation to Roxby. Information provided to the CCF is usually taken from the WMC public relations 'showbag', ranging from readily accessible annual reports to documents with colourful covers presenting the results of corporate self-monitoring.

Obviously all community-based groups have to decide whether or not they wish to be involved in governmental and corporate round tables. After some discussion, CCSA decided to participate in these negotiations on a temporary basis. The validity of participation was constantly re-evaluated. One of the reasons why conservation stakeholders finally withdrew from the process in April 1999, after a year, was that the government representatives rarely spoke at the 'round tables', leaving this task to corporate public relations executives. The 'last straw' for the two environmental representatives came when they sent a list of written concerns to the 'independent chair', Stephen

Kevin Buzzacott, an elder of the Arabunna people, who is prosecuting a case against WMC managing director Hugh Morgan, accusing him of genocide against the Arabunna.
Phil Bradley

Walsh QC. This formal letter was responded to by WMC's public relations officer Richard Yeeles, on behalf of the chair. It was clear from this that the 'independent position' of the Olympic Dam CCF was most obviously the corporate position. There can only be advantages in participation if groups recognise that their usefulness is limited, and if this process is pursued along with a range of other options. Unfortunately, the Australian environment movement devotes vast amounts of its limited time and resources to these forums.

Although not quite so enthusiastically swallowing the ideology of wise use as WWF, many other NGOs have also entered the more socially conservative world of the marketplace. Doyle and McEachern (1998) write that:

> The appeal-to-elites method of lobbying governments is merely replaced with appeals-to-corporate-elites … These more free-market interpretations of environmental reform are culminating in some NGOs no longer seeing themselves as non-profit organisations, but as 'players' who can trade freely in the market place.

The recent emergence in Australia of Earthshare is of interest in this context. This company, originating in the USA, is in its early stages of development in Australia. Its first priority is to act as a fundraising arm for mainstream environmental organisations. These organisations, along with Earthshare representatives, contact corporations, then seek permission to approach corporation employees and propose a payroll deduction scheme. Employees of an Earthshare corporation who choose to participate, contribute part of their salary to Earthshare, which then distributes funds to member organisations. In Earthshare literature, the benefits for the company becoming a signatory are as follows:

> Wherever it has been introduced, Earthshare has been widely supported by employees. It is simple to administer (it works in the same way as any other payroll deduction) and it allows you to contribute to dozens of organisations yet deal with just one. Earthshare generates both employee and community goodwill for your organisation. (Earthshare Australia 1998)

The positive aspect of programs such as Earthshare is that they seek to overcome the reliance of NGOs on government funding. The danger, of course, relates to what controls are in place to protect the trading of what the Earthshare brochure refers to as 'goodwill'. Only so much goodwill can be sold before it is seen as a commodity, rather than an indication of integrity. It is therefore essential that selection of Earthshare corporations be carried out carefully. Early on Earthshare Australia placed its operations in imminent danger when it altered its selection processes to place the onus of proof on an objector to a

corporation's inclusion into the program, rather than on the nomina-
tor of that corporation. This was done chiefly for reasons of administrative efficiency. Under this revised procedure, numerous questionable
banks and corporations have been included on the list, which belittles
the only asset Earthshare and its alliance NGOs were trading on: their
goodwill and credibility. In Earthshare's defence, these may simply be
teething problems. These changes to the funding culture in Australian
green NGOs are immense, and some techniques are untried. At worst,
however, they are interpreted as further methods of greenwashing the
activities of big capital. Certainly, more recent steps to set up an affinity card system with one of the wealthiest companies on the planet—
American Express—are misguided.

Apart from dealing in corporate board rooms and at shareholder
meetings, the new private enterprise mentality also holds the view that
nature can be 'saved' by wilderness networks through the purchase of
real estate. The Australian Bush Heritage Fund is an excellent example.
The *Australian* reported: 'An estimated 4000 Australians now pour
$750,000 into the fund each year ... The Fund yesterday announced
its seventh property acquisition, 120 ha of unprotected coastal heath
land between the Freycinet and Friendly Beaches national parks on
Tasmania's east coast' (Montgomery 1997). In the absence of government intervention (a free-market term for state activity), however, such
purchases will increase through a sense of desperation.

In October 1997 a 'brand positioning and optimisation study' was
prepared for by the ACF by AMR: Quantum Harris (1997, p 2). The
'market research' was considered necessary due to the fact that the
ACF had a steady decline in membership of approximately 6 per cent
per year, and the organisation 'sought to assess its position in the marketplace relative to its competitors'. Both Greenpeace and WWF, for
different reasons, were considered more ideally suited to the current
'market'. In its summary of findings, AMR: Quantum Harris (1997,
p 11) gave the 'ideal' in marketing terms of a major environmental
organisation:

- In a general market context, none of the major environmental organisations currently fit with the expectations of the 'ideal' environmental organisation.
- Primary qualities of the ideal organisation include: professionalism
  and working with industry and business.
- Secondary qualities include: 'friendly and approachable', 'links with
  the Government' and 'being effective'.

The study concluded that no significant barrier prevents either
the ACF or TWS assuming this 'ideal'. These findings are hardly surprising, as market research is inherently based in the conceptual

frameworks of business, as all ideas are merely considered 'commodities' that can be bought and sold. Marketable 'commodities' are most usually those that are more easily sold to the majority. Some of the more hardline, uncompromising arguments of the ACF would not be appropriate in this context. Friends of the Earth does not even rate a mention in this survey, as FoE's raison d'être has little to do with 'marketability'. If environmental organisations continue to waste large sums of money employing marketing organisations to advise them that they should become marketable, then long-term environmental change will become impossible. Obviously, there are major trade-offs in any market-based successes.

WILDERNESS CAMPAIGN VALUES CHANGE: BUILDING 'THE GLOBAL SOUL'

When Australian environmental groups do manage to look beyond domestic politics (and this is very rare), their understanding of what has to be done in the global marketplace is vastly different to that of many of their counterparts in less wealthy nations. Southern movements (widely acknowledged in green movements as the new leaders of social movement politics) have far more sophisticated strategies, a clear sense of centredness and of who their allies are. The dissolution of the concept of the 'other' has not occurred in the South: the boundaries have often become clearer with the acknowledgement that globalisation has delivered disproportionate amounts of wealth and power to Northern and Southern elites (see Doyle 1999 in press). Unfortunately, the Australian movement's perception of the effects of globalisation is quite limited.

Many mainstream environment groups in Australia perceive the environmental problems of globalisation as being, in large part, satisfactorily addressed by the creation and provision of an agreed upon, global set of moral, ethical and spiritual principles. Within this framework the Earth is now understood as a system of atomised individuals. The development of the Earth Charter deserves attention here. The Earth Charter, now being drafted, arose from the Rio Earth Summit of 1992. Maurice Strong, who championed the corporatist concept of sustainable development at the summit, is also pushing the charter, along with philanthropic capitalists from the Rockefeller Foundation. At a meeting in Canberra in February 1999,[4] Strong and Stephen Rockefeller argued before Australian representatives of mainstream environmental, church, indigenous and social justice organisations that technology (within this discourse technology is deemed value-neutral) has globalised the planet. In a bid to control the global market (which has emerged as a result of this technology) delegates were urged to

pledge their support for the establishment of a global charter enunci-
ating the shared spiritual, moral and ethical principles that, Strong and
Rockefeller argued, we all share. Delegates were urged to make a commitment to sell this global charter as a means of providing the market
(capitalism is rarely mentioned) with a 'global soul'. Such a soul will
control the natural forces of the market, which is not based, it was
assumed, on any ethical or political framework.

In his speech to the Canberra forum, Rockefeller spoke of a 'mil-
lennial campaign' that would deliver 'civilisational security' and a
'global civil society'. Rockefellar argued, as he constantly quoted from
the King James version of the Bible, that the campaign would succeed
if the Earth's people agreed through consensus to these shared moral
principles. The preamble to the Earth Charter reads: 'As a global civilisation comes into being, we can renew efforts to build a truly democratic world ... These aspirations can only be fulfilled if each and every
person acquires an awareness of global interdependence and decides to
live according to a sense of universal responsibility' (Earth Charter
Drafting Team 1999).

The seven-person charter drafting team based in the USA under
Rockefellar's guidance constantly advocates a particular type of diversity. The second general principal in its current draft reads: 'Care for
earth's community of life in all its diversity' (Earth Charter Drafting
Team 1999). While advocating this diversity, it is of some concern that
the charter does away with terms denoting groups, such as race, class,
gender, culture and species. Rockefeller (1999) defends this decision
on the grounds that 'this is not consensus language'. Instead, life on
Earth is portrayed in a neo-liberal fashion, as a series of atomised, individual beings who all share essentialist visions of the future. This attack
on community and cultural difference is also part of a much broader
business response against notions of affirmative action. Carol Bacchi
(1999) writes:

> A new equity discourse 'managing diversity' is rapidly surplanting 'equal
> opportunity' in Western democracies ... I locate managing diversity in
> part in business responses to affirmative action ... Initially, in the United
> States, Canada and Australia the business reaction to government-regu-
> lated affirmative action was hostility. The reform was described as unjus-
> tified interference in what should be a free market.

This view of the Earth comprising a series of interchangeable, individualised human parts fits in neatly with the market. In the words of
Maurice Strong at the Earth Charter meeting, 'the Earth is a corporation'. All beings are seen as consumers and providers, all systems of collective action that may generate opposition are done away with because
'we are all just people, just global citizens'. Rockefeller referred to this

new global geopolitics as 'postmodern'. The delegates too readily acknowledged this interpretation and the prescriptive pathways that 'logically followed'. The suggestion is that in this 'postmodern milieu' the salvation of the Earth will come from within individuals, not the politics of communities.

Nature is further commodified within this framework. The Earth Charter, which champions the principles of sustainable development and wise use, is committed to 'global values change', while denying any power differentials between cultures and people. Furthermore, the unwritten text of this process is that there is something wrong with the values of the majority world. Environmental problems are seen as emerging from the 'incorrect' value systems of the South, rather than from the pressures on local communities to remain competitive in the new, globalised free market. The Earth Charter is largely a Northern document, full of cultural imperialism.

Mainstream wilderness organisations such as TWS and the ACF have not backed a winner here. No doubt there are some monetary and cross-campaign trade-offs with mainstream United States-based organisations also involved in this process, but by their continued involvement, wilderness groups will increasingly be regarded as elitist and irrelevant to the majority of Australians, let alone to most inhabitants of their ill-conceived notion of a postmodern planet.

MINORITY ACTIONS AND MASS MOBILISATION CAMPAIGNS

Some environmentalists have returned to the mass-mobilisation and grassroots campaigns of the 1960s and 1970s. These strategies, however, still comprise a subservient tradition. As we saw in chapters 7 and 8, genuine mass-mobilisation methods in Australia are usually more radical than appeal-to-elites ones. Rather than accepting the corporate model, whether through lobbying corporate elites or by working closely with them, these activists work outside the state and outside business-as-usual. Consequently, they are often more conscious and critical of the models of politics that surround them, and seek more fundamental systemic changes, challenging the existence and continued power of the elites.

Although this world view can be supported by elitist/structuralist understandings of power and the state, during this third period, revolutionary change is not part of the program. Mass protests occur, predominantly in major cities and at the sites of 'threatened environments'. These are promoted and advertised as public events. They are usually one-off, with very little grassroots organising occurring beforehand, and with almost no follow-up to the event. They are almost always non-violent. In the Australian definition of mass mobilisation (as distinct from the Philippines experience, for example), people are not mobilised to provide an alternative force, but rather to seek changes in individual and collective behaviour.

Current forms of mass mobilisation are more short-term and disjointed than similar mobilisations of the earlier period. They also attract fewer people. One explanation is that people are now jaded by actions such as street marches, which have lost their spark of originality since the early days of anti-Vietnam and anti-nuclear mass demonstrations in the 1960s and 1970s. Malloch and Fulcher (1998, p 9) write:

> In the Melbourne student movement over recent years there has been a tendency to denigrate the possibility, or even desirability, of mass street mobilisations ... Comments such as 'Mass rallies are just passive strolls through the city that don't do anything', and 'Only by occupying and shutting down business as usual can we win' have crept into the anti-Jabiluka campaign in Melbourne.

One avenue activists have pursued due to this dwindling support for sustained mass actions is 'minority actions', which are forms of direct action (see chapter 3), but on a more local level than the mass rally. They usually involve a small group or network of coordinated people and include the occupation of buildings and creative stunts for media consumption. In the case of Jabiluka, network suggestions for activists included mass account openings and closings in order to tie up Westpac branches for several hours or the whole day. Although they have always been part of the environment movement's political picture, these actions are on the rise in Australian green politics.

Another reason why large-scale attempts at invoking grassroots support is becoming more difficult may be that, since the first period in the 1960s, the notion of 'community' has been roundly savaged by government policy, coupled with the dominance of the globalisation agenda. Now that the concept of a society of citizens has been undervalued and replaced with a postmodern market of separate consumers, the question emerges: 'How can there be mass mobilisation when there is no community?' The links between community and the possibility of mass actions are fairly obvious.

One outcome of this attack on community has been the disconnection between community groups and major environmental organisations. These organisations are now lobbied by communities at a time of crisis, rather than being genuinely driven by their needs. When local communities do coalesce over a particular issue or crisis, they are derogatorily referred to as NIMBYs (not-in-my-backyard) by business, government, and large green organisations. Surely to gather together on issues that affect the place where a community lives is a positive thing in a democracy?

Due to these demographic and policy changes of the third period, there have also been some different, but creative and contextually appropriate forms of mass action. These are far more sophisticated than the earlier style, which used petitions and mass rallies. There is still hope that through the informal politics of social movements change can be generated on a broad community scale, at a local and global level. A prime example of this direction was the 1997 Public Inquiry into Uranium initiated by the Nuclear Issues Coalition in Adelaide (see Doyle and Matthews 1998). I quote from the report of the inquiry:

> A more creative practice is found in the re-emergence of broad-based community coalitions ... The Public Inquiry was sponsored by a vast array of groups and organisations, including minor political parties, church groups, women's groups, peace groups, as well as a plethora of environmental interests. These broad coalitions will be a flourishing reality as Australia reaches the turn of the millennium, with so many citizens searching for new ways to contribute to a polity increasingly bereft of governmental access and leadership. (Public Inquiry into Uranium 1997)

The public inquiry was sponsored by a vast array of groups and organisations, including minor political parties, church groups, women's groups, peace groups and a plethora of environmental interests. In the case of the public inquiry, the community sector set up its own round table. Thus a community alliance chooses to mimic the inquiry mechanism of the state in its bid to pursue its own goals. All 'stakeholders' are invited to take part, but on this occasion the agenda and minimal rules for entry have been established by environmentalists and other community activists.

The inquiry proved to be a tremendous educational tool, as information about the uranium industry remains highly inaccessible. The inquiry focused largely on WMC's operation at Roxby. It gathered evidence on the industry's operation in six main areas: health and safety, uranium tailings management, information access, water management, and its contribution to the nuclear fuel cycle (Public Inquiry into Uranium 1997). In addition, it promoted solidarity among community groups. This is important, as it is often the only power resource these participants possess. It also proved to be an excellent form of theatre, promoting good media coverage. As a supplement, old hoary issues of information and scientific 'objectivity', the usefulness of adversarial processes, and numerous ethical issues also re-emerged into the public realm. The Public Inquiry into Uranium was radical, was part of a longer term struggle and was involved in mass education and symbolic politics. During the third period, however, there is less emphasis in mobilisation campaigns on the development of broad-based solidarity, and more on pursuing a variety of pathways.

The ideology of wise use has pushed the environment movement into confusion, to the edge of survival, but its frontal attack has given its younger and more radical networks new life. The movement seeks new and creative ways of playing environmental politics, some working with and against business and other sectors. Its palimpsest form, with disparate centres of power, gives it certain strengths and weaknesses. In many ways, its form, when invoked as a model for action, presents a dilemma. On the one hand, given the current political climate, it provides resilience, and a series of creative ways in which to pursue change outside politics-as-usual. Thus what could be understood as its poststructural form is its major strength. Lakshman Yapa (1999, p 17) supports this strategic model of power when he writes:

> Power is employed and exercised through a net-like organisation, individuals are not only its consenting target, but they are also the elements of its articulation. Power is in play in small individual parts and it is exercised from innumerable points in concrete actions. There are creative productive points of resistance everywhere in the power network. There is 'no single locus of great Refusal,' no soul of revolt, no grand projects that are global or revolutionary.

On the other hand is the reality of global capitalism. Can a series of uncoordinated episodes of resistance, however varied, tackle this dysfunctional giant whose existence much postmodern analysis denies? As Yapa puts it, 'Non-sovereign power is not exclusively possessed by a dominant class, or the state, there is no great divide between rulers and ruled.' In the nations of the North (such as Australia) there are not many other options at the moment, with all oppositional and dissenting voices either scattered or obliterated, and environmental activists increasingly bypassing the state. In many ways, capitalism in Australia has won—for now. In the nations of the South, with voices of dissent still being heard, and with alternative systems of meaning on which to build these voices, more direct conflicts with the 'other' must ensue, with the prescriptive models of Northern poststructuralism providing few answers but, rather, the prospect of absolute disempowerment.

Has postmodernity returned us to the early positions of the pluralists? If there is 'no soul of revolt', we might ask what can challenge the 'global soul' of Earth as corporation.

NOTES

1   There is much support within the industry sector for what has been termed 'ISO 14000' (and other permutations). The ISO documentation is a wonderful example of this assertion on the part of business for its right to self-regulate.
2   The uranium and copper mine at Roxby Downs is undergoing an enormous expansion. It is increasing production from 85,000 tonnes of ore per year to 150,000 tonnes. Its ultimate projected output is 200,000 to 350,000 tonnes.

    During this expansion phase, WMC has obtained permission to increase its extraction of fossil waters from the Great Artesian Basin from 14 megalitres per day (Ml/d) to 42 Ml/d.

3   I served as one of two 'conservation stakeholders' on the ODCCF. This account is ethnographic by nature, cross-referenced by more verifiable official and unofficial documentation.

4   As President of the CCSA I was invited to participate at this forum in Canberra, February–March 1999. This account is based on my notes of the meeting.

CONCLUSION: REPOPULATING THE ENVIRONMENT

Until as recently as five years ago, the environment movement in Australia reflected a particularly narrow definition of 'the environment'. Two key trends emerged. In the more radical groups, 'wilderness' agendas dominated, with their concomitant 'ecocentric' arguments about the rights of 'other nature'. Other groups—the resource conservationists—treated the environment as some form of instrumental resource, to be cajoled and juggled into something more malleable, more easily used by humans. Both traditional approaches perceive the 'environment' as 'somewhere out there', not usually including people.

More recently, the environment movement in Australia has been challenged to reshape its agenda into something more reminiscent of broad-ranging green movements elsewhere. The resource conservationists are still there, but are operating under a subtle change of nomenclature, now referred to as 'sustainable developers', 'environmental managers', 'wise use planners' and 'biodiversity programmers'. The more radical agenda is now being shaped by 'social ecological' concerns: the relationships between people, and between people and other species, are now deemed important. Reflecting this change in the social movement's

agenda is the inclusion of issues relating to urban and rural environments, domestic and workplace environments, the rights of indigenous
peoples, other issues of social equity, non-violence and democracy, while
still advocating the rights of the non-human (see Doyle 1994b).

On the urban front much imaginative work by Australian environmentalists is under way. The work of Peter Newman and Jeff Kenworthy
is of note here, as they grapple with transforming cities (where most of
us live) into a more vibrant, urban form (Newman 1995). Adelaide's
Ecopolis development, pioneered by Paul Downton, Cherie Hoyle and
Urban Ecology Australia Inc., has revealed an important element of
this new environmental radicalism: its emphasis on the Dreaming.
Inspirational architectural and decision-making models for society are
being concocted without the self-censorship of the reformists. The
Ecopolis development in Adelaide seeks to reconstruct one city block
in a way that promotes community and diversity, while providing part
of its own energy, and recycling its own waste (Orszanski 1994). There
are practical restrictions at the implementation level, but these should
not be utmost in people's minds at the level of imagination.

Many good environmental programs are being put into place in
established communities. These programs do not seek to create a new
city, but to recognise the community-generated life and knowledge
already evident in parts of the city (Darley 1994). At long last we begin
to recognise that less affluent people have been suffering environmental problems for generations: problems of health, the problems of
everyday survival that arise through living next to noxious and haz-
ardous trades and arterial traffic routes. These problems do not need
to be monitored and measured by the creation of sophisticated data
bases: the local people know they exist.

In many ways, ecopolises once existed in Australian cities—before,
for example, the creation of multinational shopping centres that neces-
sitated the rerouting of public transport systems, resulting in community centres becoming peripheral. The need to conserve resources is
still of utmost importance, but notions of social equity and redistribu-
tion are foremost on this new environmental agenda. With the rein-
clusion of the human in nature, other environments previously
ignored—the domestic and workplace environments—also come into
play. The work of Elaine Stratford is interesting. An academic by training, Stratford (1994) has written quite extensively on the place of her
own home and others, in an attempt to radicalise the home as a site for
research and action in environmental studies.

Australian environmentalists must do more to address issues pertaining to indigenous peoples. For too long 'culture' has been excluded from the symbol 'environment'. Wilderness is a non-Aboriginal
construct that sees nature/the environment as hostile, wild, romantic

and, most importantly, removed from humans. Aboriginal people see themselves as inseparable from nature. While partly sharing this view, non-Aboriginal people have often used this perspective to perpetuate their perceived racial superiority. Consequently, indigenous peoples are seen as part of the fauna and flora, and their more holistic connections with nature are ignored (Bayet 1994). It was encouraging to witness the Australian Conservation Foundation's wholehearted involvement in the Wik debate and its informal support of the Jabiluka Action Group. Hopefully this will be a benchmark for other environmental organisations in the future.

In the first chapter of this book the myth of the common goal was discussed at length. The environment movement in Australia is a true

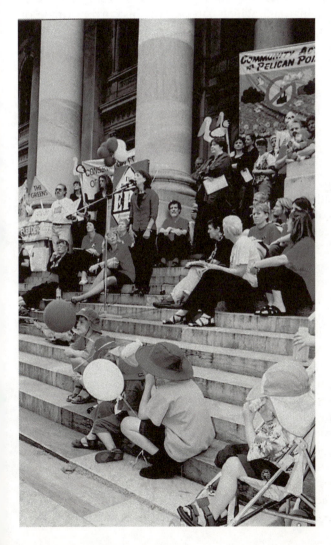

Concerns about environmental justice finally emerge in the Australian environment movement.
Aussie Kanck

social movement. It is diverse, amorphous and flexible, and clusters around loose notions of symbolic identity. If we accept my description of this teleologically non-unified phenomenon, it is tempting simply to conclude that as the movement is multifaceted, its prescriptive and strategic approaches will follow suit. It is critical, however, to recapitulate the distinction here between 'description' and 'prescription': these purposes are not the same.

There may, on occasions, be some prescriptive advantages of the palimpsest form of the politics of dissent, most particularly in the context of the North. This concept allows the movement to survive, to become resilient and pursue a diversity of approaches, at a time when direct confrontation and resistance often mean disempowerment and alienation. Some argue that an ecopolitics of resilience (using some elements of postmodern analysis), which can outlast this current tirade of laissez-faire economics and ecology, can only occur when strategic alliances are made in all parts of society, within and outside the business community, the state, the union movement, and other social movements.

Yet some activists have used this concept to justify neo-liberal responses, to go with 'the flow', to deal with anyone, regardless of moral or political position, and to reject the more disagreeable voices from within. In this manner, in the EPBC negotiations it was easy for the government to find green activists willing to flag its narrow position, while excluding those opposed.

Apart from the difference between the purposes of prescription and description, there is also the fundamental issue of context. The former position, for example, can be prescriptively appropriate when it combines the palimpsest model of action with an understanding of political context: I refer to this position as the 'Trojan horse' argument. The latter position, on the other hand, does not consider that there are qualitative distinctions between political contexts that allow decisions and strategies to be informed ones, rather than an ad hoc pursuit of everything and nothing at the same time: we might call this position 'the universal embrace'.

Both arguments are based on the assumption that the environment movement cannot return to the exclusively outsider politics of the first period, despite the tough conditions of the third period, but the conclusions drawn in relation to strategic practice are very different. This shared assumption is mostly valid, and may explain, in part, why the two positions are often confused. The Trojan horse argument then asserts that environmentalists must deal with 'the other' in a bid to make that position more 'green'. In short, environmentalists must stop exclusively preaching to 'the converted', and actively coopt the positions of business and the state. At the same time the Trojan horse position acknowledges that the power of the movement is not often comparable to the power of business, either in quantity, the way it is structured, or the way it manifests itself. These power discrepancies are most sorely felt when groups attend business-led forums. Thus there will be times for engagement with 'the other', and times for withdrawal. In certain circumstances, for example, valuable information about political adversaries can be gained through cooperation. On other occasions, there are even chances of genuine environmental reform. Conversely, when a round table is evidently a public relations exercise designed for quieting opposition (as is the case of the Olympic Dam Community Consultative Forum), there may be a need to withdraw. Finally, in other instances, a decision may have to be made simply not to engage in the first place, focusing on grassroots activities outside politics-as-usual. These decisions are based on theory and context.

The 'universal embrace' position is strategically useless, and dangerously sells out the power of movement politics. It is more jelly-like and relativistic, arguing that the concept of 'the other' is limiting. With notions of the other disappearing, as Rockefeller and Strong's Earth Charter insists, we are now told that we are all just global citizens, wanting the same things: 'we are all just "living beings", aren't we'. Power differentials between people and groups are no longer included in this extreme version of postmodern reality. This has led to a situation where many activists are willing to sit down in an illusory, unidimensional, 'level playing field' and negotiate with any grouping, regardless of political and moral context. This is the misguided political style of the professionals of those organisations involved in the EPBC negotiations.

On many occasions, alliances with particular industry, government and other sectors must be pursued; on others, it remains entirely inappropriate and disempowering. The movement, profoundly anti-intellectual (again reflecting its society), often incorrectly interprets this postmodern political space as a justification for strategies that are simply neo-liberal and ultimately deliver market-based outcomes.

This neo-liberalism currently pervades the operational and strategic reality of the environment movement in Australia, particularly in elite networks: 'Find your own comfort level in social and political systems and pursue your own environmental ends and ecotopia will be revealed.' But not all roads lead to ecotopia (whatever that is), not all political processes are equally useful in achieving solutions to environmental crises (whatever they are). Many directly conflicting and ambiguous objectives rally under the flag of 'environment'. They cannot and should not all be universally endorsed. Some processes, some strategies, some goals are more usefully embarked upon to solve different environmental problems, given the limited resources of the environment movement, and some work to counter any notion of a 'green society'.

Can we simply say that not all positions are given equal weight, that the arguments of the powerful are more persuasive than those of the weak? The extreme neo-liberalism of some environmentalists has not dealt with the basic problems of access to power and wealth. Partly as a consequence, the movement has become dominated, at a structural level, by professional, career bureaucrats and corporate players with extremely limited objectives. The more radical, alternative messages of the movement have ceased to be heard. So the task remains not necessarily to essentialise but to rediscover some of these more radical messages, to highlight some of the more radical traditions that have emerged over the past three decades, to critique these traditions and to reshape what we mean by environmentalism at the beginning of the twenty-first century. Some of these traditions are non-violent, anti-bureaucratic, pro-democracy and empowerment-oriented, multidirectional, ecumenical, biocentric, socially just, interconnected and symbolic. Other social movements may share some of these values and characteristics, but not all.

Any effective use of postmodern/poststructural strategies must also include adversarial, structuralist, front-on, us-versus-them, win-loss conflicts, as well as other forms of appeals-to-elites, win-win, and mass-mobilisation politics. In their acceptance of 'universal embrace' strategies, many Australian environmental strategists are overindulging the characteristic of postmodern politics that White (1991) alludes to: in their 'irreducible pluralism and a suspicion of totalistic revolutionary programs' they rarely offer combined and confrontational stances, which are sometimes most appropriate in creative forms of postmodern politics. Instead, what is eulogised as postmodern cleverness usually manifests itself as neo-pluralism and neo-liberalism.

There are some *descriptive* advantages of postmodern and poststructural models of power. There are many weaknesses in these arguments if they are used to *prescribe* future actions that question the power of capital as a class, most obviously manifested in the operations of large corporations. Some 'new philosophers' have used Foucault's work to discount the possibility of any real change in the distribution of power among people (see Anderson 1988). Let us not forget as we juggle the subtle strengths and weaknesses of structural and poststructural perceptions of the deployment of power, that most people in the North operate everyday under the notion that the pluralist system of power is not a model at all, but simply their reality. The pluralist democratic system holds that every citizen has equal access to power, with no additional power in the hands of any specific interest group or section of society.

It does not accept that large corporations have a privileged position towards the state (Lindblom 1977), let alone that they often dictate terms, or that they are sometimes indistinguishable from the state.

If we partly accept that, as Stephen Bell (1995) argues, 'the whole world has now become a capitalist system ... with its capacity to terrorise and discipline national governments through currency depreciation or capital flight', then we must not be afraid to view the interests of large corporations and business interests as, if not the same, then very similar. At the same time, we must not overemphasise their unity. Their power can also be quite fragile (Simpson 1995).

One of the strengths of postmodern analysis is that it has opened up possibilities for understanding the world in more ways than just simple dualisms. This has allowed social movements to leave entrenched positions, where adversaries are always depicted as 'the forces of darkness'. But this position leads to tactical weaknesses when all notions of identity are lost in an overly relativistic political outlook. Depictions of power that do away with all notions of 'the other' (even at a symbolic level) may actually reaffirm capitalism's dominance, as more oppositional conflicts are deemed 'passé'.

The fact that the movement does not have a rational, common goal need not undermine the notion of collective identity or action. Identity, created and sustained through myriad networks of power, *is* the movement. It is still necessary, therefore, to understand the movement in terms of 'we'. Wendy Wheeler (1994, p 105) contends:

> Part of the challenge which a radical and emancipatory ... democratic politics now faces is, first, that of acknowledging a plurality within reason and identity, second, that of understanding that popular expressions are symptomatic, and third, that of finding ways of making a political response to the nostalgic desire for communalities. This means articulating a politics capable of constituting a 'we' which is not essentialist, fixed, separatist, divisive, defensive or exclusive. Clearly, the structures and forms of such a politics must be capable of representing both the diversity and the desire of contemporary identities.

This is the challenge of movement politics in the early years of the new millennium.

The Australian movement has overly indulged in domestic politics. When it has entered international forums it has been largely on the basis of issues defined as appropriate by domestic party-political agendas, such as global warming. It is hoped that the movement will become further involved in global issues, understanding that many Australian environmental problems are also experienced elsewhere and, in the countries of the South, often in a very direct way.

The pressures of predominantly Northern-driven globalisation on the planet and its people are critical. There are few better examples of the free-market economic solutions peddled by the North than that of Indonesia. The dramatic crash of the Indonesian rupiah, which lost 75 per cent of its value in 1997–98, led to a range of 'bail-out' initiatives proposed by a coalition of US and International Monetary Fund interests. Rather than focusing on the urgent restoration of the value of the rupiah, the IMF promised fiscal assistance, but at the same time demanded further 'liberalisation' of the local market. The Indonesian economy and its environment suffered under the long dictatorship of Suharto, with all its associated cronyism and nepotism. What had further contributed to the Indonesian crisis, however, was the earlier globalisation and 'freeing up' of its market. Among numerous outcomes, this process led to vast outbreaks of currency speculation by predominantly Northern traders, resulting in outrageously high inflation and unemployment. Technological developments (for example, electronic or virtual currency trading) facilitating the rapid transfer of capital between nation-states had weakened Indonesia's capacity to regulate its economic affairs. By denying these effects and by concentrating on the 'negatives' of the Indonesian economy and culture, the IMF has further exacerbated the 'Asian crisis'. What is worse, however, is the renewed vigour with which the IMF is pushing for further deregulation of Indonesian society in the wake of the 'crisis'. Among these proposals is a reduction by two-thirds of the public sector, condemning many essential environmental infrastructure projects to the scrap heap. In the words of IMF head Michel Camdessus (1998), 'all restrictive marketing arrangements will be abolished, leaving firms free to produce and export their products as they wish, and as the market decides'.

Brian Martin (1993) argues that the most significant change in power relations as a result of this privatisation of policy 'has not been from public to private, nor from the state to the market, but from national and local political agendas to global centres of economic power'. This leads us to the Multilateral Agreement on Investment (MAI). From 1995 to 1998, 29 of the Earth's wealthiest nations, which comprise the Organisation for Economic Cooperation and Development (OECD), attempted to forge a Multilateral Agreement on Investment. The MAI was designed to bind and limit the power of nation-states (and their citizens) to impose their own requirements, whether they be financial, environmental or human rights-based, on multinational investors. If this agreement had been signed by the OECD (and imposed on the South), any national legislation setting a minimum wage for its workforce, or minimum environmental standards, would have been deemed a restraint on trade. Nation-states could then be sued on this basis. The MAI attempted to provide unrestricted rights for transnational companies to:

- export their commodities or services without conditions;
- own any ecosystem or resource without any obligations;
- accumulate profits within any nation without responsibility for reinvestment;
- access government grants normally provided to domestic operators; and
- invest in any privatised public infrastructure.

Marjorie Griffen Cohen (1997, p 6) says of the MAI:

> Unlike the work of nation states, which over time have developed institutions either to correct the economy when the market did not function in an optimal way (such as during times of depression), or to control business, such as through labour or environmental legislation, the international replacements that are being created neither exert discipline on the market nor function as instruments of market correction. The new international institutions are designed solely to discipline nations in the interests of one class which enjoys world citizenship—the international investor.

Contrary to popular perceptions, the MAI is not dead, but is re-emerging in a range of other mechanisms. In mid-1999, negotiations over forest resources under the 'takings legislation' of the World Trade Organisation is one just example of national legislation being watered down to allow unregulated market transactions for transnational logging interests.

So where do environmentalists go from here? Well, certainly not to Rio, nor any of the Brundtland/UNCED-inspired conferences since Rio. Modern nation-states are limited in their ability to monitor and regulate transnational corporate activities. With the emergence of the Northern agenda of global ecology, many environmental issues are seen as beyond the traditional scope of national governments. Governments often lag behind in their responses, 'and this transnational political space has been occupied by corporations and NGOs, which can cross nation-state boundaries more readily. This globalisation of ecological and market systems has led to "the politics of no fixed address"' (Doyle and McEachern 1998, p 105). Jacques Attali (1991, pp 5–6), who served as the foundational head of the European Bank for Reconstruction and Development, looked into the near future:

> Severed from any national allegiance or family ties by microchip-based gadgets that will enable individuals to carry out for themselves many of the functions of health, education, and security, the consumer-citizens of the world's privileged regions will become 'rich nomads.' Able to participate in the liberal market culture of political and economic choice, they will roam the planet seeking ways to use their free time, shopping for

information, sensations, and goods only they can afford, while yearning for human fellowship, and the certitudes of home and community that no longer exist because their functions have become obsolete. Like New Yorkers who every day face homeless beggars who loiter around automated teller machines pleading for spare change, these wealthy wanderers will everywhere be confronted by roving masses of 'poor nomads'—boat people on a planetary scale—seeking to escape from the destitute periphery, where most of the earth's population will continue to live. These impoverished migrants will ply the planet, searching for sustenance and shelter, their desires inflamed by the ubiquitous and seductive images of consumerism they will see on satellite TV broadcasts from Paris, Los Angeles, or Tokyo. Desperately hoping to shift from what Alvin Toffler has called the slow world to the fast world, they will live the life of the living dead.

This is the world of wise and sequential use, of the gang bang theory of nature. The forces that seem most capable of moving beyond the boundaries of nation-states in hot pursuit of transnational corporations are social movements and NGOs, also acting through transnational conduits. At first glance, the age-old story depicting the battle between David and Goliath seems an appropriate metaphor to describe this situation. Only this time, David has no sling-shot.[1]

An interconnected movement is one that understands the importance of its informal networks, how they are structured, and how they function. The Australian environment movement needs to further nationalise and globalise its networks, sharing resources and information derived from other bioregions and countries. The first part of this book describes and celebrates the diverse web of networks—with their diverse purposes—that make up the environment movement. The networks include formal and informal relations cutting across traditional sectors. More can be done to promote, to service and to cherish what these networks already do.

In this book much has been made of the characteristics of different political forms. Certain structures are best used to promote specific types of green objectives. Informal networks and groups can continue to play more radical, imaginative and confrontational brands of ecopolitics. More formal environmental NGOs must accept that these more informal players have an incredibly important role, and should not only be tolerated but nourished.

Green NGOs must use their links with universities, governments, political parties, businesses, unions, and other community groups. Flexibility is important. As ecological and social contexts change, so must political structures adapt. Both adaptability and amorphousness relate to flexibility: to be ever-changing, continually growing, and to be perpetually learning and experimenting in differing political contexts.

Pursuing these networks, however, is not an excuse to bypass the state. Postmodern analyses that promote this course will be found to be flawed in the medium and long terms. Without a moderate, representative and liberal/social democratic state to legislate against the excesses, crimes and misdemeanours of an antisocial minority, acute environmental degradation will increase. Industry, as well as other vested interests, cannot be left to self-regulate. Their 'win-win' equations are a nonsense. Governments, however hostile, must be continually engaged. The doors of elected officials and public servants must be knocked on, and if unanswered, kicked down. All regimes end. If we deny national and state governments their key role, it will be impossible to implement longer term administrative systems necessary for human existence on the Earth.

Community, in the Filipino and majority world sense, still includes close associations over a period of time between people, as well as a connection, often sacred, to a *place*. In Australia, in both urban and rural areas, this connection to place has dissipated, and an affiliation with people over time has been made increasingly difficult with rapid demographic changes. Consequently, community organising in Australia means something completely different to the Filipino perspective: attempts at dabbling with mass media, and raising support for one-off marches and rallies. In Australia we seek to reconstruct notions of community for a brief moment of protest.

It would be foolish to blame the dissolution of community in Australia on postwar demography. The impact of globalisation, which allows companies such as WMC to enter lands once owned by the local indigenous people in another country, has also further exacerbated the situation. We are now told, at least in the wealthier nations of the North, that we are part of a global community, and that past associations are increasingly irrelevant. This attempt to replace communities of peoples and places with a megasphere of individualised commodities of natural and human products does not create a global community at all, but a non-community, a society bereft of local systems of meaning, local biophysical support systems and limitations, and senses of identity and purpose that unite them.

The metanarrative of global, hyper-capitalism, which quickens the market and its fiscal transactions by using supply chain maximisation and other cyber products, cannot tolerate national or local boundaries that define communities of identity and impede its profit-making potential. In Australia, these pressures to denounce local associations and to disempower dissenting voices has even led to the bastardisation of the very concept of community. The government and business interests now fund 'community organisations' formed to service government grants schemes (largely funded from the ongoing sale of Telstra).

Such groups as Landcare, Coastcare, Tidy Towns, Bushcare, Our Patch, and Local Agenda 21 associations are cheap, on-the-ground service providers for these government schemes and programs. These groups are closely controlled and harnessed by bureaucratic and industry initiatives and arrangements. Applying the term 'community' to these service providers has the effect of disempowering grassroots community activists. On many occasions, these groups do valuable work. But they are *not* community groups and organisations as defined in this book. They are not part of the tradition of the modern environment movement that genuinely emerged from community politics.

The environment movement in Australia emerged from the community, in the non-institutional networks of our lives. Its continued health depends on the health of its communities. To an extent, environmental action can assist people who have resettled in a new place,

Local people, often referred to as 'NIMBYS', are increasingly taking up the role of environmental watchdogs across Australia.
Aussie Kanck

among new people, to reconstitute a community. Environmental issues often provide a focus for the reinvigoration of the community, to give it a raison d'être. Communities can only be re-created, or strengthened, with interactions between people. Without a shared politics, there is no community, without community, there is no politics.

Under the wise use ideology of the Howard government, NIMBY (not-in-my-backyard) activists are regarded with suspicion. But NIMBYs are part of an environmental tradition in this country that gives voice to people, to communities interested in the *place* where they live, work and recreate. NIMBYs are to be valued, for their very existence challenges the neo-pluralist and market-based notions of our polity that regard humans in their 'natural state' as apolitical. Ignore the celebratory publications heralding our arrival in Ecotopia because of the emergence of limited, incremental environmental legislation, and a slight adjustment to business practices. The silos are now green, orchids are protected under the shadow of aluminium smelters, and we now have half-flush and full-flush toilets.

Despite the rhetoric, the accommodation, the cooption and now the rejection, there remains in Australia a vibrant, alternative political voice known as the environment movement. It still has much to do.

NOTES

1   I first used this phrase in an article entitled 'Sustainable Development and Agenda 21: the Secular Bible of Global Free Markets and Pluralist Democracy', *Third World Quarterly*, vol. 19, no. 4, 1998, pp 771–86.

REFERENCES

Abrahams, H (1987), letter to K Alexander, 13 August.

Abrahams, H, Bond, D and Wilson, L (1987), 'Evaluation Queensland', report prepared for Melbourne evaluation meeting, 13 July.

ACF (Australian Conservation Foundation), (1978, 1982), *The Green Pages: Directory of Non-Government Environmental Groups in Australia*, ACF, Melbourne.

——(1996), '77th Meeting of the ACF Council', 10–12 August 1996, ACF, Fitzroy, Vic.

Adams, D (1980), *The Hitch Hiker's Guide to the Galaxy: a Trilogy in Four Parts*, Pan, London.

Advertiser (1987), 16 June, p 10.

Alexander, K (1987), 'Election Evaluation, July 1987', draft, Melbourne.

AMR: Quantum Harris (1997), 'Brand Positioning and Optimisation Study', Melbourne.

Anderson, J (1984), *Public Policy-Making*, 3rd edn, Holt, Rinehart & Winston, New York.

Anderson, R (1988), *The Power and the Word: Language, Power and Change*, Paladin Grafton Books, London.

Anon. (1993), 'Letter to the Editor', *Entelechy* (University of Adelaide), pp 2–3.

Anon. (1996), 'Letter to the Editor', *Megazine* (University of Adelaide), p 1.

Attali, J (1991), *Millennium: Winners and Losers in the Coming World Order*, Times Books, New York.

Australian Agency for International Development (1998), 'Indonesia Australia Specialised Training Project', advertisement in the *Australian*, 4–5 April.

Australian Nature Conservation Agency (1997), 'The Ramsar Convention', brochure.

Australian Transpersonal Conference (1986), 'advertising notice'.

Axelrod, R (1984), *The Evolution of Cooperation*, Basic Books, New York.

Bacchi, C (1999), 'Managing Diversity: A Contested Concept', *The International Review of Women and Leadership*, December.

Bachrach, P and Baratz, M (1962), 'The Two Faces of Power', *American Political Science Review*, 56, pp 947–52.

Baker, S (1997), letter to author, 5 June.

——(1999), letter to author, 8 August.

Bayet, F (1994), 'Overturning the Doctrine: Indigenous people and Wilderness—Being Aboriginal in the Environment Movement', *Social Alternatives*, vol. 13, no. 2, pp 27–32.

Beder, S (1993), *The Nature of Sustainable Development*, Scribe, Newham, Vic.

——(1996), *Global Spin*, Scribe Publications, Newham, Vic.

Bell, S (1987), 'Socialism and Ecology: Will Ever the Twain Meet?', *Social Alternatives*, 6, 3, pp 5–12.

——(1995), 'The Environment—a Fly in the Ointment', *Chain Reaction*, no. 73–74, pp 30–3.

Bita, N (1995), 'Up a Gumtree: The Greens in Disarray', *Australian*, 20 December.

Bookchin, M (1980), 'Ecology and Revolutionary Thought,' in *Environment and Society*, eds RT Roelofs, JN Crowley, and DL Hardesty, Prentice-Hall, Englewood Cliffs, NJ, pp 187–94.

——(1982), *The Ecology of Freedom: the Emergence and Dissolution of Hierarchy*, Cheshire Books, Palo Alto, Calif.

Boreham, G and Milburn, C (1995), 'Caswell Latest Casualty of Greens Disarray', *Age*, 6 May, p 19.

Bottomore, TB (1977), *Elites & Society*, Penguin, Harmondsworth.

Boudin, R (1988), 'Common Sense and the Human Sciences', *International Sociology*, vol. 3, no. 1, March.

Bridgeman, P and Davis, G (1998), *Australian Policy Handbook*, Allen & Unwin, Sydney.

Brinkehoff, MB and Kunz, PR (1972), *Complex Organizations and their Environments*, Wm. C. Brown Company Publishers, Iowa.

Britell, J (1997), 'Partnerships, Roundtables and Quincy-Type Groups are Bad Ideas that Cannot Resolve Environmental Conflicts', at <http://www.harbourside.com/home/j/jbritell/welcome.htm>.

Broad, R (1994), 'The Poor and the Environment: Friends or Foes?', *World Development*, 22, 6, pp 811–22.

Brow, A (1997), Green Evolution: The Evolution of Australia's National Environmental Pressure Groups in the 1990s, Minor Dissertation, MEnvSt, Mawson Graduate Centre for Environmental Studies, University of Adelaide.

Bujra, JM (1973), 'The Dynamics of Political Action: A New Look at Factionalism', *American Anthropologist*, vol. 75, no. 1, February, pp 132–52.

Bury, JB (1960), *The Idea of Progress*, Dover Publications, New York.

CAFNEC (Cairns and Far North Environment Centre) (1984), 'Cape Tribulation to Bloomfield Road Blockade Information Sheet for Site Occupiers and Support Personnel', January.

Camdessus, M (1998), 'Statement by the Managing Director on the IMF Program with Indonesia', IMF, Washington DC, news brief no. 98/2, 15 January.

Cant, S (1993), 'Democrats Demand Minister Stop Logging,' *Australian*, 6 December, p 2.

Capra, F (1982), *The Turning Point: Science, Society, and the Rising Culture*, Simon & Schuster, New York.

Carson, R (1962), *Silent Spring*, Penguin Books, Harmondsworth.

Cawson, A and Saunders, P (1981), 'Corporatism, Competitive Politics and Class Struggle', paper presented at BSA/PSA Conference on Capital, Ideology and Politics, quoted in C Ham and M Hill (1993), *The Policy Process in the Modern Capitalist State*, 2nd edn, Harvester Wheatsheaf, New York and London.

Chandler, J (1992), 'Green Guerillas', *Age*, 18 January.

Chapin, FS and Tsouderos, JE (1956), 'The Formalization Process in Voluntary Organizations', *Social Forces*, vol. 34, no. 4, pp 342–4.

Chatterjee, P and Finger, M (1994), *The Earth Brokers: Power, Politics and World Development*, Routledge, London and New York.

Claessen, HJM and Skalnik, P (eds) (1978), *The Early State*, Mouton, The Hague.

Clinton, B, quoted in Dowie, M (1995), *Losing Ground: American Environmentalism at the Close of the Twentieth Century*, MIT Press, Cambridge, Mass. and London.

Cohen, I and Seed, J (1984), 'In Defence of Daintree,' *World Rainforest Report,* no. 2, August, a newsletter produced by the Rainforest Information Centre, Lismore, NSW.

Commoner, B (1971), *The Closing Circle,* Knopf, New York.

Connor, J (1999), 'ACF EPBC', *Habitat Australia,* August 1999.

CCSA (Conservation Council of South Australia) (1998a), pamphlet, CCSA, Adelaide.

——(1998b), press release, 20 April.

Cotgrove, S and Duff, A (1980), 'Environmentalism, Middle-Class Radicalism and Politics', *The Sociological Review,* New Series, vol. 28, no. 2, May, pp 333–51.

Courier-Mail (1986), 16 May, p 3.

Crick, B (1962), *In Defence of Politics,* Weidenfeld & Nicolson, London.

Cuthbertson, GM (1975), *Political Myth and Epic,* Michigan State University Press, Michigan.

Dahl, R (1969), 'The Concept of Power', in *Political Power: A Reader in Theory and Research,* eds R Bell, D Edwards, R Harrison Wagner, Free Press, New York.

Dardick, G (1986), 'Gary Snyder: of Place and the Buddha-realm,' *Simply Living,* vol. 2, no. 11, pp 38–9.

Darley, S (1994), 'But the Working Class Don't Care About the Environment … Do They?', *Social Alternatives* (University of Queensland, Brisbane), vol. 13, no. 2, pp 37–41.

——(1996), letter to the author, 7 August.

Davis, B (1981), Characteristics and Influence of the Australian Conservation Movement: An Examination of Selected Conservation Controversies, PhD thesis, Department of Government, University of Tasmania.

Davies, J (1971), *When Men Revolt and Why,* Free Press, London and New York.

Devall, B (1981), 'The Deep Ecology Movement', *Natural Resources Journal,* vol. 20, pp 229–321.

Devall, B and Sessions, G (1985), *Deep Ecology: Living as if Nature Mattered,* Peregrine Smith Books, Salt Lake City Utah.

Diani, M (1992), 'The Concept of Social Movement', *The Sociological Review,* vol. 40, no. 1, February, pp 1–25.

Dickens, P (1990), *Urban Sociology: Society, Locality and Human Nature,* Harvester Wheatsheaf, London.

Donati, PR (1984), 'Organisation Between Movement and Institution', *Social Science Information,* New Delhi, 23, 4/5, pp 837–59.

Dowie, M (1995), *Losing Ground: American Environmentalism at the Close of the Twentieth Century,* MIT Press, Boston, Mass.

Down to Earth, Queensland Newsletter, April 1986.

Downer, A (1987), 'ACF Election Decision No Surprise', press release, 15 June.

Downes, D (1996), 'Neo-Corporatism and Environmental Policy', *Australian Journal of Political Science,* 31 (2), pp 175–90.

Doyle, T (1986), 'The Structure of the Conservation Movement in Queensland', *Social Alternatives,* vol. 5, no. 2, pp 27–32.

——(1987a), 'Environmental Politics in Queensland: the Lindeman Island Affair', *The Journal of the Royal Historical Society of Queensland,* vol. 13, pp 462–72.

——(1987b), 'The Myth of the Common Goal: the Conservation Movement in Queensland', *Social Alternatives,* vol. 6, no. 4, pp 33–6.

——(1991), 'The Green Elite and the 1987 Election', *Chain Reaction,* no. 63–64, pp 26–31.

——(1992a), 'Dissent and the Environment Movement', *Social Alternatives,* vol. 11, no. 1, pp 24–6.

——(1992b), 'Mass Media and Popular Culture: A Business or an Expression of Our Lives?', *Social Alternatives,* vol. 11, no. 1, pp 3–7.

——(1994a), 'Direct Action in Environmental Conflict in Australia: A Re-examination of Non-Violent Action', *Regional Journal of Social Issues,* no. 28, pp 1–13.

——(1994b), 'Radical Environmentalism: Repopulating the Environment', *Social Alternatives*, vol 13, no. 2, pp 5–7.

——(1995), 'Corporations, Power and the Environment', *Chain Reaction*, no. 73–74, pp 14–17.

——(1998a), 'Sustainable Development and Agenda 21: The Secular Bible of Global Free Markets and Pluralist Democracy', *Third World Quarterly*, University of London, vol. 19(4), pp 771–86.

——(1998b), 'Green Politics in Grey Times: the Politics of Resilience—The Phil Tighe Memorial Lecture', *Ecopolitics XI Conference Proceedings*, University of Melbourne, Melbourne, pp v–xvi.

——(1998c), 'Western Mining Corporation (Australia) and the Olympic Dam Community Consultative Forum: Playing "Wise Use" Roundtable Games', paper presented at the International Conference on Mining Transnational Corporations, at the Palm Plaza Hotel, Manila, November 14–16.

——(1999), 'Roundtable Decision Making in Arid Lands under Conservative Governments', in *Environmental Policy Two*, eds KJ Walker and K Crowley, University of New South Wales Press, Sydney, pp 122–41.

——(in press), *Environmental Movements*, Routledge, London and New York.

——(forthcoming), 'The Campaign to Save the Wet Tropical Forests', in *Securing the Wet Tropics: Environmental Management in Queensland*, eds M Lane and G McDonald, Federation Press, Sydney.

Doyle, T and Kellow, A (1995), *Environmental Politics and Policy Making in Australia*, Macmillan, Melbourne.

Doyle, T and Matthews, D (1998), 'Playing Green Politics Outside the State: The Public Inquiry into Uranium', *Ecopolitics XI Conference Proceedings*, University of Melbourne, Melbourne.

Doyle, T and McEachern, D (1998), *Environment and Politics*, Routledge, London and New York.

Doyle, T and Walker, K (1996), 'Looking for a World They Can Call Their Own', *Campus Review*, vol. 6, no. 3, February, pp 1, 7.

Drengson, AR, 'Shifting Paradigms: From Technocratic to the Person Planetary,' *Environmental Ethics*, vol. 3, 1980, pp 221–40.

Dryzek, J (1983), 'Ecological Rationality', *International Journal of Environmental Studies*, vol. 21, pp 5–10.

——(1987), *Rational Ecology: Environment and Political Economy*, Basil Blackwell, Oxford and New York.

——(1997), *The Politics of the Earth: Environmental Discourses*, Oxford University Press, London and New York.

Duverger, M (1972), *Party Politics and Pressure Groups: A Comparative Introduction*, Nelson, London.

Dye, T and Zeigler, LH (1970), *The Irony of Democracy: an Uncommon Introduction to American Politics*, Wadsworth Pub. Co., Belmont, Calif.

Earth Charter Drafting Team (1999), 'Earth Charter General Principles Draft 1999', Canberra, February.

Earthshare Australia (1998), 'What is Earthshare?', adverstising pamphlet, Bangalow, NSW.

Eckersley, R (1992), *Environmentalism and Political Theory: Toward an Ecocentric Approach*, UCL Press, London.

——(1993), 'Free Market Environmentalism: Friend or Foe?', *Environmental Politics*, 2, 1, pp 1–19.

Economou, N (1992), 'Problems in Environmental Policy Creation: Tasmania's Wesley Vale Pulp Mill Dispute', in *Australian Environmental Policy*, ed K Walker, University of New South Wales Press, Sydney.

——(1993), 'Accordism and the Environment: the Resource Assessment Commission and National Environmental Policy-Making', *Australian Journal of Political Science*, vol. 28, no. 3, November, pp 399–412.

Edelman, M (1964), *The Symbolic Uses of Politics*, University of Illinois Press, Urbana, Ill.

Ekins, P (1992), *A New World Order: Grassroots Movements for Global Change*, Routledge, London.

Elix, J, 'Forest Campaign has Significant Election Effect', *ACF Newletter*, vol. 19, no. 7, p 2.

Environmental Defenders Office, NSW (1999), 'EDO NSW Analysis of the Environment Protection and Biodiversity Conservation Act', September.

Ehrlich, P (1968), *The Population Bomb*, Ballantine Books, New York.

Etzioni, A (1968), *The Active Society*, Basic Books, New York.

Faich, RG and Gale, RP (1971), 'The Environmental Movement: From Recreation to Politics', *Pacific Sociological Review*, vol. 14, no. 3, July, pp 270–87.

Fiske, J (1993), *Power Plays, Power Works*, Verso, London and New York.

Flew, A (1970), 'Introduction', in T Malthus, *An Essay on the Principle of Population*, ed A Flew, Penguin Books, Harmondsworth, pp 17–48.

FoE (Friends of the Earth) (1975), 'Politics of the Middle Class', *Chain Reaction: Bulletin of Friends of the Earth Australia*, summer edn.

——(undated), 'Starting a Friends of the Earth Group', pamphlet, FoE, Fitzroy, Vic.

Foreman, D, and Haywood, B (eds) (1987), *Ecodefense: A Field Guide to Monkeywrenching*, 2nd edn, Ned Ludd Books, Tucson, Arizona.

Foucault, M (1976), *The History of Sexuality*, trans. Robert Hurley, Allen Lane, London.

Fox, W (1990), *Toward a Transpersonal Ecology: Developing New Foundations for Environmentalism*, Shambala, Boston, Mass.

Frohlich, N and Oppenheimer, JA (1978), *Modern Political Economy*, Prentice-Hall, Englewood Cliffs, NJ.

Gaventa, J (1980), *Power and Powerlessness: Quiescence and Rebellion in an Appalachian Valley*, University of Illinois Press, Chicago.

Gerlach, LP and Hine, VH (1970), *People, Power, Change: Movements of Social Transformation*, Bobbs-Merrill Educational Publishing, Indianapolis.

Gerth, H and Mills, C (1946), *From Max Weber: Essays in Sociology*, Oxford University Press, New York.

Ghai, D and Vivian, JM (eds) (1992), *Grassroots Environmental Action: People's Participation in Sustainable Development*, Routledge, London and New York.

Godfrey-Smith, W, 'Environmental Philosophy: Are You a "Shallow Environmentalist" or a "Deep Ecologist"', *Habitat*, 1979, pp 24–5.

Golding, W (1955), *Lord of the Flies*, Coward-McCann, New York.

Gouldner, AW (1959), 'Organizational Analysis', in *Sociology Today: Problems and Prospects*, eds RK Merton, L Broom and LS Cotrell, Jr, Basic Books, New York, pp 400–10.

Gowan, S, Lakey, G, Moyer, W and Taylor, R (1976), *Moving Towards a New Society*, New Society Press, Philadelphia.

Grady, M (1998), 'SA's Biggest Park—Juggernaut in Low Gear', *Environment South Australia*, vol. 7, no. 3, p 13.

Grainger, A (1980), 'A Hymn to the Forest', pamphlet.

Green, J (1998), 'Australia's Anti-nuclear Movement', *Green Left Weekly*, 26 August, pp 14–15.

Green, R (1984), 'Greenies Act On Joh's Land', *Sydney Morning Herald*, 7 January, p 26.

Greenpeace (1996a), '1995 Annual Report', *Greenpeace Australia News* 6 (2), pp 11–14.

——(1996b), 'Corporate Actions may save North Sea fish stocks', *Greenpeace Business*, London, Greenpeace, 31, June–July.

Griffin Cohen, M (1997), 'Presentation to the House of Commons Sub-Committee on International Trade, Trade Disputes and Investment, Hearing on the MAI Panel on Corporate, Consumer and Social Implications', Ottawa, Canada, at <http://www.policyalternatives.ca/maipresentation.html>.

Hall, C (1987), letter to the author, 19 July.

Hall, S and du Gay, P (eds) (1996), *Questions of Cultural Identity*, Sage, London.

Ham, C and Hill, M (1993), *The Policy Process in the Modern Capitalist State*, 2nd edn, Harvester Wheatsheaf, New York and London.

'Hans' (1984), quoted in the Douglas Shire's Wilderness Action Group, *The Trials of Tribulation*, p 24.

Hare, B (1987), '1987 Federal Election: Summary of Discussions with TWS, National Meeting on 29/5/87, at Sydney—TWS', May 31.

Harris, C (1995), 'Comments on Direct Action', received by The Wilderness Society, 9 March.

Harry, J, Gale, R and Hendee, J (1969), 'Conservation: An Upper-Middle Class Social Movement', *Journal of Leisure Research*, vol. 1, pp 246–54.

Hartshorn, GS (1991), 'Key Environmental Issues for Developing Countries', *Journal of International Affairs*, vol. 7, pp 393–401.

Hay, P (1987), 'Land Use Politics in Tasmania', *Current Affairs Bulletin*, vol. 64, no. 3, pp 4–12.

——(1991), 'Destabilising Tasmanian Politics: The Key Role of the Greens', *Bulletin of the Centre for Tasmanian Historical Studies*, vol. 3, no. 2, pp 60–70.

——(1994), 'The Politics of Tasmania's World Heritage Area: Contesting the Democratic Subject', *Environmental Politics*, vol. 3, no 1, spring.

Hay, P and Haward, M (1988), 'Comparative Green Politics: Beyond the European Context?', *Political Studies*, vol. 36, pp 433–48.

Held, D (1987), *Models of Democracy*, Stanford University Press, Stanford, Calif.

Helvarg, D (1994), *The War Against the Greens: The 'Wise Use' Movement, the New Right, and Anti-Environmental Violence*, Sierra Club Books, San Francisco, Calif.

Henderson, J (1985), interview with the author, 3 August.

Higgins, E (1987), 'The Taint of Politics Splits the Greenies', *Times on Sunday*, 18 October, p 13.

Hill, Robert (1984), quoted in *Sydney Morning Herald*, 7 January, p 26.

——(1998a), letter to peak environment groups, September.

——(1998b), quoted in ABC Radio Interview, Port Lincoln, August.

Hill, Rosemary (CAFNEC) (1985), as quoted by the author at the 'Special General Meeting of the Queensland Conservation Council', 27 July.

Hobbes, T (1914), *Leviathan*, introd. AD Lindsay, Dutton, New York.

Holloway, G (1986), 'The Social Composition of the Environment Movement in Australia', Paper presented at the Ecopolitics Conference, Brisbane, 30–31 August.

——(1987), letter to the author.

——(1990), Access to Power: the Organisational Structure of the Wilderness Conservation and Anti-Nuclear Movements in Australia, PhD dissertation, Department of Sociology, University of Tasmania.

Hughes, V (1998), letter to R Playfair, 17 September.

Hundloe, T (1985), 'The Environment', in *The Bjelke-Petersen Premiership 1968–1983: Isues in Public Policy*, ed AP Patience, Longman Cheshire, Melbourne.

Hutton, D (1987), *Green Politics in Australa*, Angus & Robertson, Sydney and London.

Hutton, D and Connors, L (1999), *A History of the Australian Environment Movement*, Cambridge University Press, Cambridge, UK.

Imura, H (1994), 'Japan's Environmental Balancing Act', *Asian Survey* XXXIV, 4, pp 355–68.

Inglehart, R (1977), *The Silent Revolution*, Princeton University Press, Princeton, NJ.

International Monetary Fund (IMF), (1994) cited in R Broad, 'The Poor and the Environment: Friends or Foes?', in *World Development*, 22, 6, pp 811–22.

Jones, C (1987), letter to J West, 2 May.

Judah, S (1974), *Hare Krishna and the Counterculture*, Wiley, New York.

Jung, CG (1938), *Psychology and Religion*, Yale University Press, New Haven.

Kant, I (1984), *Idea of a Universal History on a Cosmopolitan Plan*, trans. with an introduction and notes by SL Jaki, Edinburgh Scottish Academic Press, Edinburgh.

Kelly, P (1994), *The End of Certainty: Power, Politics and Business in Australia*, Allen & Unwin, St Leonards, NSW.

Kerin, R (1998), letter to CCSA, 24 April.

Kiernan, K (1990), 'I Saw my Temple Ransacked', in *The Rest of the World is Watching: Tasmania and the Greens*, eds C Pybus and R Flanagan, Pan Macmillan, Sydney.

Klapp, O (1972), *Currents of Unrest*, Holt, Rinehart & Winston, New York.

Klandermans, B, Kriesi, H and Tarrow, S (eds) (1989), *From Structure to Action: Comparing Social Movement Research Theory Cultures*, JAI, London.

Krishnan, K (1978), *Prophecy and Progress*, Allen Lane, London, 1978.

Kuhn, T (1962), *The Structure of Scientific Revolutions*, University of Chicago Press, Chicago and London.

Lambert, G (1987), 'Federal Elections and Their Opportunities for TWS', pamphlet, TWS, Sydney.

Lambert, J (1994), 'Campaigning for Wilderness in the '90s', *Wilderness News*, August–September, p 10.

Law, G (1996), 'Litany of Environmental Crimes by the Coalition Government' (draft).

Lawrence, DH (1989), *Kangaroo*, Collins Publishers, Sydney.

Leahy, P and Mazur, A (1978), 'Comparison of Movements Opposed to Nuclear Power, Fluoridation, and Abortion', *Research in Social Movements, Conflict and Change* no. 1, pp 143–54.

Lindblom, C (1977), *Politics and Markets: The World's Political-Economic System*, Basic Books, New York.

LETS South Australia (1997), advertising brochure, Adelaide.

Lofland, J and Stark, R (1965), 'Becoming a world-saver: A theory of conversion to a deviant perspective', *American Sociological Review*, 30, pp 862–74.

Lowe, P and Goyder, J (1983), *Environmental Groups in Politics*, George Allen & Unwin, London.

Lukes, S (1974), *Power: A Radical View*, Macmillan, London.

Lunn, S (1998), Greenhouse Influence Easily Bought *Australian*, 5 Feb, p 2.

Malloch, L and Fulcher, R (1998), 'Jabiluka: How Can We Win?', *Green Left Weekly*, Sydney, 9 December, p 9.

Malthus, TH (1826), *An Essay on the Principles of Population*, 6th edn, J Murray Publishers, London.

Martin, B (1980), 'The Scientific Straight-Jacket', *The Ecologist* (UK), vol. 11, no. 1 pp 33–43.

——(1984), Environmentalism and Electoralism', *The Ecologist* (UK), vol. 14, no. 3, pp 110-18.

——(1992), 'Intellectual Suppression: Why Environmental Scientists are Afraid to Speak Out', *Habitat*, vol. 20, no. 3, pp 11–14.

——(1993), *In the Public Interest?: Privatisation and Public Sector Reform*, Zed Books, London.

Mausner, B (1979), *A Citizen's Guide to the Social Sciences*, Nelson-Hall, Chicago.

Mazur, N (1995), 'Who Pays in the Zoo', *Chain Reaction*, no. 73–74, May, pp 34–7.

Mazza, P (1997), 'Cooptation or Constructive Engagement?: Quincy Library Group's Efforts to Bring Together Loggers and Environmentalists Under Fire,' *Cascadia Planet*, at <http://www.cascadia@tnews.com>, August.

McEachern, D (1991), *Business Mates: The Power and Politics of the Hawke Era*, Prentice-Hall, Sydney.

——(1993), 'Environmental Policy in Australia 1981–91: A Form of Corporatism?', *Australian Journal of Public Administration*, vol. 52, no. 2, pp 173–86.

——(1995), 'Mining Companies and the Defence of Nature', *Chain Reaction*, no. 73–74, May, pp 18–21.

McGee, 'Mr' (1984), 'Far North Queensland Scene,' *Mareeba Advertiser*, 15 August.

McNicol, J, Murray, J and Leeks, R (1985), *Unite*, a leaflet of the Brisbane Non-Violent Action Training Collective.

Melucci, A (1984), 'An End to Social Movements? Introductory Paper to the Sessions on "New Movements and Change in Organisational Forms", *Social Science Information*, 23, 4/5, pp 819–35.

Merchant, C (1993), *Radical Ecology: the Search for a Livable World*, Routledge, New York.

Meredith, P (1999), *Myles and Milo*, Allen & Unwin, St Leonards, NSW.

Metcalf, W (1984), 'A Classification of Alternative Lifestyle Groups,' *Australian and New Zealand Journal of Sociology*, vol. 20, no. 1, March.

Michels, R (1911, 1962), *Political Parties*, Free Press, New York.

Middlemas, K (1979), *Politics in Industrial Society*, Andre Deutsch, London.

Minerals Policy Institute (1998), *Glossy Reports, Grim Reality: Examining the Gap between a Mining Company's Social and Environmental Record and its Public Relations Campaigns*, MPI, Sydney.

Montgomery, B (1994), 'Thesis Claims "Defame"', *Australian*, 1 June, p 21.

——(1997), 'Bush Fund Buys Seventh Wonder', *Australian*, 10 December.

Moore, A (1986) 'Pagan Festival Group', *News From Home* (newsletter of the Australian Association for Sustainable Communities), vol. 3, May.

Moore, P (1997), 'Consensus,' *Greenspirit*, at <http://www.pmoore@rogers.wave.ca>.

Moorhead, G, Ference, R and Neck, CP (1987), 'Group Decision Fiascos Continue: Space Shuttle Challenger and a Revised Groupthink Framework', *Human Relations Journal*, vol. 44, no. 6, p 547.

Morgan, H, quoted in McEachern, D (1991), *Business Mates: The Power and Politics of the Hawke Era*, Prentice-Hall, Sydney, pp 15–19.

Morgan-Thomas, E (1987), letter to J West, 15 April.

Morris, A and Herring, C (1987), 'Theory and Research in Social Movements: A Critical Review', *Annual Review of Political Science*, vol. 2, pp 137–98.

Mosca, G (1939), *The Ruling Class*, trans. H Kahn, MacGraw Hill, London.

Mosley, JG (1987), letter to the author.

Moxham, N, Baker S et al. (1996), 'Report on Peak Council/National Environmental Consultative Forum to Friends of the Earth Australia', November, Friends of the Earth, Fitzroy, Vic.

Moyo, S, O'Keefe, P and Middleton, N (1993), *The Tears of the Crocodile: From Rio to Reality in the Developing World*, Pluto Press, London.

Naess, A (1972), 'The Shallow and the Deep Ecology Movements', a summary of an introductory lecture at the Third World Future Research Conference, Bucharest, 3–10 September.

Neidhardt, F and Rucht, D (1990), 'The Analysis of Social Movements: the State of the Art and Some Perspectives for Future Research', in *Research in Social Movements: The State of the Art*, ed D Rucht, Campus/Boulder Co., Westview Press, Frankfurt.

Newman, K (1980), 'Incipient Bureaucracy: the Development of Hierarchies in Egalitarian Organizations', in *Hierarchy and Society: Anthropological Perspectives on Bureaucracy*, Institute for the Study of Human Issues, Philadelphia, pp 143–64.

Newman, P (1995), 'Ecologically Sustainable Cities: Alternative Models and Urban Mythology', *Social Alternatives*, vol. 13, no. 2, pp 13–18.

Niewenhausen, H (1985), interview with the author, 3 May.

Nordlinger, E (1981), *On the Autonomy of the Democratic State*, Harvard University Press, Cambridge, Mass.

Notion, H (1991), 'Greenpeace, Getting a Piece of the Green Action', *Chain Reaction*, no. 63–64, pp 32–4.

Oakeshott, M (1962), *Rationalism in Politics, and Other Essays*, Basic Books, New York.

Obershall, A (1976), *Social Conflict and Social Movements*, Prentice Hall, New York.

Olsen, M (1970), *Power in Society*, Macmillan, New York.

Olson, M (1965), *The Logic of Collective Action*, Harvard University Press, Cambridge, Mass.

O'Neill, J (1994), 'Greenpeace Runs Out of Steam', *Independent Monthly*, July.

O'Riordan, T (1976), *Environmentalism*, Pion Limited, London.

Orszanski, R (1994), 'The Halifax Ecocity Project', *Social Alternatives*, vol. 13, no. 2, pp 33–6.

Ostrom, E (1990), *Governing the Commons: the Evolution of Institutions for Collective Action*, Cambridge University Press, Cambridge and New York.

Ostrom, E, Burger, J, Field, C, Norgaard, R and Policansky, D (1999), 'Revisiting the Commons: Local Lessons, Global Challenges', *Science*, vol. 284, April, pp 278–82.

Pakulski, J (1991), *Social Movements: the Politics of Moral Protest*, Longman Cheshire, Melbourne.

Panitch, L (1980), 'Recent Theorisations of Corporatism: Reflections on the Growth Industry', *British Journal of Sociology*, 31 (2).

Papadakis, E (1993), *Politics and the Environment: The Australian Experience*, Allen & Unwin, Sydney.

——(1996), *Environmental Politics and Institutional Change*, Cambridge University Press, New York.

——(1998), *Historical Dictionary of the Green Movement*, Scarecrow Press, Maryland.

Parker, K (1996), 'From the Campaign Coordinator', *Wilderness News*, 144, autumn, p 3.

Perrow, C (1970), *Organizational Analysis: A Sociological View*, Tavistock Publications, UK.

Pfeffer, J (1981), *Power in Organizations*, Graduate School of Business, Stanford University, & Pitman Publishing Inc, Mass.

Piven, FF and Cloward, RA (1979), *Poor People's Movements: Why They Succeed. How They Fail*, Vintage, New York.

Plamenatz, J (1973), *Democracy and Illusion*, Longmans, London.

Po-Keung Ip (1978), 'Taoism and the Foundations of Environmental Ethics', *Environmental Ethics*, vol. 5, pp 335–43.

Poulantzas, N (1973), 'The Problem of the Capitalist State', in *Power in Britain*, eds J Urry and J Wakeford, Heinemann, London.

Prideaux, M, Horstman, M and Emmett, J (1998), 'Sustainable Use or Multiple Abuse,' *Habitat*, vol. 26, no. 2, April, p 15.

Princen, T and Finger, M (1994), *Environmental NGOs in World Politics: Linking the Local to the Global*, Routledge, London and New York.

Public Inquiry into Uranium (1997) (T Doyle, chair), *Report of the Public Inquiry into Uranium*, CCSA, Adelaide, November.

Pybus, C and Flanagan, R (eds) (1990), *The Rest of the World is Watching: Tasmania and the Greens*, Pan Macmillan, Sydney.

Rae, M (1987a), letter to the author, 20 August.

——(1987b), 'The Wilderness Society's Plan for 1987/88. What Do We Do And How Do We Do It?', TWS, Melbourne.

Rainbow, S (1992), 'Why did New Zealand and Tasmania spawn the world's first green parties?', Occasional Paper 23, Centre for Environmental Studies, University of Tasmania.

Redclift, M (1987), *Sustainable Development: Exploring the Constructions*, Methuen, London.

——(1994), 'Sustainable Development: Economics and the Environment', in *Strategies for Sustainable Development: Local Agendas for the Southern Hemisphere*, eds M Redclift and C Sage, John Wiley and Sons, London, pp 17–34.

RMP (Resource Monitoring and Planning Pty Ltd) (1998), letter to the author, 17 August.

Richards, L (1997), *Western Mining in the Philippines*, Indigenous Philippines Press, Sydney.

Rochford Jr, EB (1982), 'Recruitment Strategies, Ideology, and Organization in the Hare Krishna Movement', *Social Problems*, vol. 29, no. 4, April, pp 399–410.

Rockefeller, S (1999), presentation at the Earth Charter seminar, Canberra, February.

Rootes, C (1987), 'Social Movements—An Overview and Prospect', *Social Alternatives*, vol. 6, no. 4, November, pp 2–5.

Rosewarne, C (1995), 'Undermining Aboriginal Interests', *Chain Reaction*, no. 73–74, May, pp 38–41.

Rowell, A (1996), *Green Backlash: Global Subversion of the Environmental Movement*, Routledge, London and New York.

Rushdie, S (1989), 'The Riddle of Midnight', an Antelope Production for Channel Four, 1988, screened by SBS on 'Masterpiece', February.

Russell, B (1961), *History of Western Philosophy*, Allen & Unwin, 2nd edn, London.

Ruzicka, E (1987), 'Meet Jonathon', *Wilderness News*, vol. 8, no. 1, February.

Saint-Simon, H (1814 (1976–1985)), *The Reorganization of the European Community*, Edition du Tricentenaire, Paris.

Salleh, A (1996), 'Politics in/of the Wilderness', *Arena* 23, June–July.

Saunders, M and Summy, R (1986), *The Australian Peace Movement: A Short History*, Peace Research Centre, ANU, Canberra.

Schmitter, P (1974), 'Still the Century of Corporatism?', *Review of Politics*, vol. 36, pp 85–131.

Schreiber, H (1991), 'The Threat for Environmental Destruction in Eastern Europe', *Journal of International Affairs*, vol 3, no. 1, pp 361–91.

Seed, J, quoted in Chandler, J, 'Green Guerillas,' *Age*, 18 January.

Seeman, M (1959), 'On the Meaning of Alienation', *American Sociological Review*, 24, pp 783–91.

Seeman, M (1975), 'Alienation studies', *Review of Sociology*, 1, pp 91–122.

Shaw, BV (1996), 'Was the Wilderness Society Tricked into Supporting the Coalition at the Last Federal Election?', paper given to the Public Policy Program, ANU, Canberra.

Sibeon, R (1996), *Contemporary Sociology and Policy Analysis: the New Sociology of Public Policy*, Tudor, Eastham, Merseyside.

Sierra Club (1997), '"Local Control" a Smokescreen for Logging,' *The Planet: The Sierra Club Activist Resource*, at <http://www.sierraclub.org/planet/199711/delbert.html>.

Sills, DL (1971), *The Volunteers. Means and Ends in a National Organisation*, The Free Press, Glencoe, Ill.

Simon, D (1989), 'Sustainable Development: Theoretical Construct or Attainable Goal', *Environmental Conservation*, vol. 6, no. 1.

Simpson, A (1994), 'Ethnographic report on the Activities of EAAAC', unpublished report, Adelaide.

——(1995), 'Shoalwater Bay—a Win for the Environment Movement?', *Chain Reaction*, no. 73–74, pp 42–4.

——(1996a), letter to J Downey, 4 November

——(1996b), letter to K Parker, 9 October.

——(1996c), letter to C Walker, 10 October.

Sinclair, J (1986), 'Factions Cause Concern', Letters to the Editor, *ACF Newsletter*, vol. 18, no. 4, May.

Sison, JM and de Lima, J (1998), *Philippine Economy and Politics*, Aklatng Bayan Publishing House, Philippines.

Smith, C and Freedman, A (1972), *Voluntary Associations: Perspectives on the Literature*, Harvard University Press, Cambridge, Mass.

Smith, R (1992), 'Greenpeace: Behind the Masks', *Sunday Age*, 17 May.

Smith, R (ed) (1993), *Politics in Australia*, 2nd edn, Allen & Unwin, Sydney.

Snow, D, Zurcher Jr, L, and Ekland-Olson, S (1980), 'Social Networks and Social Movements: A Microstructural Approach to Differential Recruitment', *American Sociological Review*, vol. 85.

Stallings, RA (1983), 'Patterns of Belief in Social Movements: Clarifications from an

Analysis of Environmental Groups', *The Sociological Quarterly*, 14, autumn, pp 465–80.

Stark, R and Bambridge, WS (1980), 'Networks of Faith: Interpersonal Bonds and Recruitment to Cults and Sects', *American Journal of Sociology*, vol. 85, pp 1376–95.

Stone, D (1988), *Policy Paradox: The Art of Political Decision Making*, WW Norton and Co., New York and London.

Stratford, E (1994), 'Thoughts on the Place of Home in Environmental Studies,' *Social Alternatives*, vol. 13, no. 2, pp 19–23.

——(1995), Construction Sites: Creating the Feminine, the Home and Nature in Australian Discourses on Health, PhD dissertation, Mawson Graduate Centre for Environmental Studies, University of Adelaide.

Stringer, R (1996), 'Food, Hunger and the Environment: Painful Lessons and Misguided Policies', Depts of Geography and Environmental Studies Joint Seminar Series, University of Adelaide, 9 September.

Strong, G (1998), 'The Green Game', *Age,* 17 August, at <http://www.the-age.com.au/daily/980717/news/news22.html>.

Summy, R and Saunders, M (1986), *A History of the Peace Movement in Australia*, University of New England, Armidale, NSW.

Sylvan, R and Bennett, DH (1986), 'Deep Ecology and Green Politics,' *The Ecologist*, no. 20, April.

Tao Te Ching, quoted in Chan Weing-Tsit (1963), *A Source Book in Chinese Philosophy*, Princeton University Press, Princeton, NJ, pp 139–73.

Taplin, R (1992), 'Adversary Procedures and Expertise: the Terania Creek Inquiry', in *Australian Environmental Policy*, ed K Walker, University of New South Wales Press, Sydney, pp 156–82.

Tarrow, S (1988), 'National Politics and Collective Action: Recent Theory and Research in Western Europe and the USA', *Annual Review of Sociology*, vol. 14, pp 421–40.

Tart, CT (1975), *Transpersonal Psychologies*, Harper & Row, New York.

Taylor, B (1991), 'The Religion and Politics of Earth First!', *The Ecologist*, vol. 21, no. 6, November–December, pp 258–66.

Taylor, B, Hadsell, H, Lorentzen, L, and Scarce, R (1992–93), 'Grassroots Resistance: The Emergence of Popular-Environmental Movements in Less Affluent Countries', *Wild Earth*, winter, pp 43–9.

The Phantom (1987), interview with author at 'Skull Cave', Coorparoo, Brisbane, 10 June.

Thompson, J (1967), 'Infinity in Mathematics and Logic', *The Encyclopedia of Philosophy*, vol. 4, Macmillan & Free Press, New York.

Thorstensen, L and Harris, A (1992), 'Let's hear it from the workers', Letters to the Editor, *Chain Reaction*, no. 65, p 4.

Tighe, P (1992), 'Hydroindustrialisation and Conservation Policy in Tasmania', in *Australian Environmental Policy*, ed K Walker, University of New South Wales Press, Sydney, pp 124–55.

Tisdell, C (1988), 'Sustainable Development: Differing Perspectives of Ecologists and Economists, and Relevance', *World Development*, vol. 16, no. 3, pp 373–84.

Toffler, A (1970), *Future Shock*, Pan, London.

Touraine, A (1980), *The Voice and the Eye: An Analysis of Social Movements*, Cambridge University Press, Paris.

Tudor, H (1972), *Political Myth*, Pall Mall Press, London.

TWS (The Wilderness Society) (1983), 'Constitution', Hobart.

TWS (The Wilderness Society) (1984), 'Effects of the NAGs/Forest People's Blockade', report to the Victorian Nature Forest Action Council, March.

TWS (The Wilderness Society) (1987), 'Federal Election 1987: Summary of Results', August, Hobart.

TWS (The Wilderness Society) (1994), *Policy and Procedures Manual*, TWS, Hobart.

TWS (The Wilderness Society) (1996), 'Coalition Environment Package Marks Start of a New Relationship', press release, Wilderness Society, Melbourne.

TWSSRG (The Wilderness Society Support/Review Group) (1988), Report, Hobart.

UNDP (United Nations Development Program) (1996), Sustainable Human Development, 'Economic Growth has Failed for a Quarter of World's People, says Report Written for UN Development Programme', press release, 17 July.

Useem, B (1980), 'Solidarity Model, Breakdown Model and the Boston Anti-Bussing Movement', *American Sociological Review*, 45 (3), pp 357–69.

Vallentine, J (1987), 'A Green Peace: Beyond Disarmament', in *Green Politics in Australia*, ed D Hutton, Angus & Robertson, Sydney, pp 49–66.

——(1989), 'Defending the Fragile Planet: The Role of the Peace Activist', in *Politics of the Future: the Role of Social Movements*, eds C Jennett and R Stewart, Macmillan, South Melbourne, pp 56–75.

——(1990), 'The Politics of Environmental reform: The Greening of the Peace Movement in Australia', *Social Alternatives*, vol. 9, no. 1, pp 4–8.

Victorian Nature Forest Action Council (1986), 'Reasons for Disassociation from the Actions of the Forest People (NAGs), informal report, March.

Walker, C (1996), 'Peak Council Meeting', Canberra, 6–10 May, Friends of the Earth, Fitzroy, Vic.

——(1997), letter to author, 15 April.

——(1999), Letter to QCC.

Walker, KJ (1987), 'Major Paradigms in the Social Sciences', Griffith University course notes, Nathan, Qld.

——(1988), 'The Environmental Crisis: A Critique of Neo-Hobbesian Responses', *Polity*, vol. xxi, no. 1, fall, pp 67–81.

Walker, KJ (ed) (1992), *Australian Environmental Policy*, University of New South Wales Press, Sydney.

Ward, I (1989), *Politics One*, Macmillan, Melbourne.

Ward, I and Cook, I (1992), 'Televised Political Advertising: Media Freedom, and Democracy', *Social Alternatives*, vol. 11, no. 1, pp 21–6.

Warhurst, J (1992), 'The Australian Conservation Foundation: Twenty-Five Years of Development', *Ecopolitics V Proceedings*, University of New South Wales, Sydney.

Waud, P (1984), 'National Meeting Discussion Paper' (TWS), 24 May.

Weber, M (1958), *The Theory of Social and Economic Organization*, Free Press, Glencoe, Ill.

Wells, D (1993), 'Green Politics and Environmental Ethics: A Defence of Human Welfare Ecology', *Australian Journal of Political Science*, vol. 28, no. 3, pp 515–27.

West, J (1984), 'A Letter from the Director', *Wilderness News*, 24 May, Hobart.

—— (1985), 'The Wet Tropics: What can we expect from the Labor Government in its Second Term?', ACF Canberra Office, January.

Wheeler, W (1995), 'Nostalgia Isn't Nasty: The Postmodernization of Parliamentary Democracy', in *Altered States: Postmodernism, Politics, Culture*, ed M Perryman, Lawrence & Wishart, London, pp 94–112.

White, S (1991), *Political Theory and Postmodernism*, Cambridge University Press, Cambridge, UK.

Wilderness Action Group (1984), *The Trials of Tribulation*, Kajola, Mossman, NSW.

Wilkinson, P (1971), *Social Movements*, Praeger, New York.

Wilson, KL and Orum, AM (1976), 'Mobilising people for collective political action', *Journal of Political and Military Sociology*, 4, pp 187–202.

Woodford, J (1995), 'How Green is the Valley Now?', *Sydney Morning Herald*, 1 July, p 33.

Wooten, H, QC (1986), 'Statement by President', *ACF Newsletter*, vol. 18, no. 6, July.

World Bank (1993), 'Privatisation: the Lessons of Experience', cited in B Martin, *In the Public Interest? Privatisation and Public Sector Reform*, London, Zed Books.

Wright, J (1977), *The Coral Battleground*, Thomas Nelson Australia, West Melbourne, Vic.

Yapa, L (1999), 'A Primer on Postmodernism with a View to Understanding Poverty', Department of Geography, Pennsylvania State University, at <http://www.geog.psu/~yapa/Discorse.htm>.

Young, J (1990), *Post Environmentalism*, Belhaven Press, London.

INDEX

outsider politics 109, 138

palimpsest xxix, 19
paradigms 109
participatory democracy 132
Pearson, Noel 91
People's Environmental Protection
    Authority 34
Phantom Club 79
Philippines x, 210
Pine Gap 116
pluralism xxii, xxvii, xxxi, 6, 118–20,
    131, 147, 152
Political Parties 37
population 67
postenvironmentalism 114
postmaterialist theory 123
postmodern states 196
postmodernism xxii, 189–92, 213, 220,
    221
poststructuralism xxii, 189, 190, 220
poverty xxv, 144
power xxxi, xxxii
prescription 218
pressure groups 119
    see also interest groups
primary industries 176
primary organisations 78
professional activists/environmentalists
    161, 163
professional elite 171
professionals 11, 23, 174
profit maximisation xxxix
public health care 145
Public Inquiry into Uranium Mining
    212
public service 186

Queensland 10, 16, 24, 32, 161
Queensland Conservation Council 14,
    18, 25

race xxv
radical libertarian solutions 178
radical libertarians 66, 146
Rainforest Action Group 58
Rainforest Action Network xv
Rainforest Conservation Society of
    Queensland 24, 28
RAMSAR convention 176
Ranger Inquiry 133
Reaganomics 191
Renegade Activist Action Force 57
Resource Assessment Commission 150
resource conservation 117

Richardson, Graham 96
Rio de Janeiro Earth Summit 86, 144,
    147
Rockefeller, Stephen 208, 219
Rockefeller Foundation 208
round table 150, 153, 181, 185, 194
    decision making 183
Roxby Downs x, 116, 141, 181, 202
*Roxby Downs (Indenture Ratification
    Act) 1982* 204
Royal Institute of Architects 127
Rum Jungle 135

sand mining 152
science 178
secondary organisations 78
sequential use xxiii, 181
Sinclair, John 116, 168
Smith, Adam 68, 146
social constructionists 65
social ecology 61, 117
social equity xxv
social movement theories 3, 6
social movements xx
socialism 70
Socialist Workers' Party 136
South America 147
South Australia x
South West Tasmania 32
Spencer, Herbert 68–69
stakeholder 181, 200
Strong, Maurice 208, 219
structuralism xxii, 118, 120, 131, 139,
    151, 152
structuralist analysis 194, 210
structuralists 197
Supreme Court of NSW 129
sustainable development 61, 141–42,
    148, 185
sustainable use xxiii, 141, 178

Tasmania 32
Tasmanian Conservation Trust 14
Tasmanian Hydro-Electric Commission
    126, 130
Tasmanian Wilderness Society 33, 39
teleology xxxi, xxxiv
Terania Creek 48, 116, 129
Terania Creek Inquiry 129
Thatcherism 191
think-tanks 178
Third World xiii, xv, 144
Toffler, Alvin 114
Total Environment Centre 116
tourism 152